Henry Llewellyn Williams, Amos Alonzo Stagg

A Scientific and Practical Treatise on American Football for Schools

and Colleges

Henry Llewellyn Williams, Amos Alonzo Stagg

A Scientific and Practical Treatise on American Football for Schools and Colleges

ISBN/EAN: 9783337414566

Printed in Europe, USA, Canada, Australia, Japan

Cover: Foto ©berggeist007 / pixelio.de

More available books at **www.hansebooks.com**

A

Scientific and Practical Treatise

ON

AMERICAN FOOTBALL

FOR

𝔖chools and ℭolleges

BY

A. ALONZO STAGG

AND

HENRY L. WILLIAMS

HARTFORD, CONN.
Press of The Case, Lockwood & Brainard Company
1893

CONTENTS.

PREFACE.

THE game of football is fast becoming the national fall sport of the American youth. Among the larger eastern colleges, where it has been fostered and developed, football has now been raised to a definite science, but in the west the game is, as yet, comparatively in its infancy

The demand has been rapidly increasing among the smaller colleges and large preparatory schools from year to year for competent coachers, and it is evident that there is felt a wide-spread want for some source of definite information which shall describe the manner of executing the various evolutions, the methods of interference, and the more difficult and complicated points of the game.

It is with the desire of meeting this want so far as is possible, and with the hope of stimulating a love for the game and of raising the standard of play among the school-boys of this country, to whom the colleges and universities must look for the material out of which to construct their future elevens, that the authors have prepared this volume.

The endeavor has been made to begin with simple steps in the early development of the game and advance by gradual stages to the most difficult evolutions and scientific tactics which have been mastered up to the present day.   In working out this principle the aim throughout has been clearness and precision.

While it is the primary desire to furnish in this work a practical aid in the attainment of a higher standard of play among the preparatory schools and colleges, still it is hoped that the general public will find it an assistance to the better understanding of American football, which has come to hold such a prominent place in popular favor.   THE AUTHORS.

September 15, 1893.

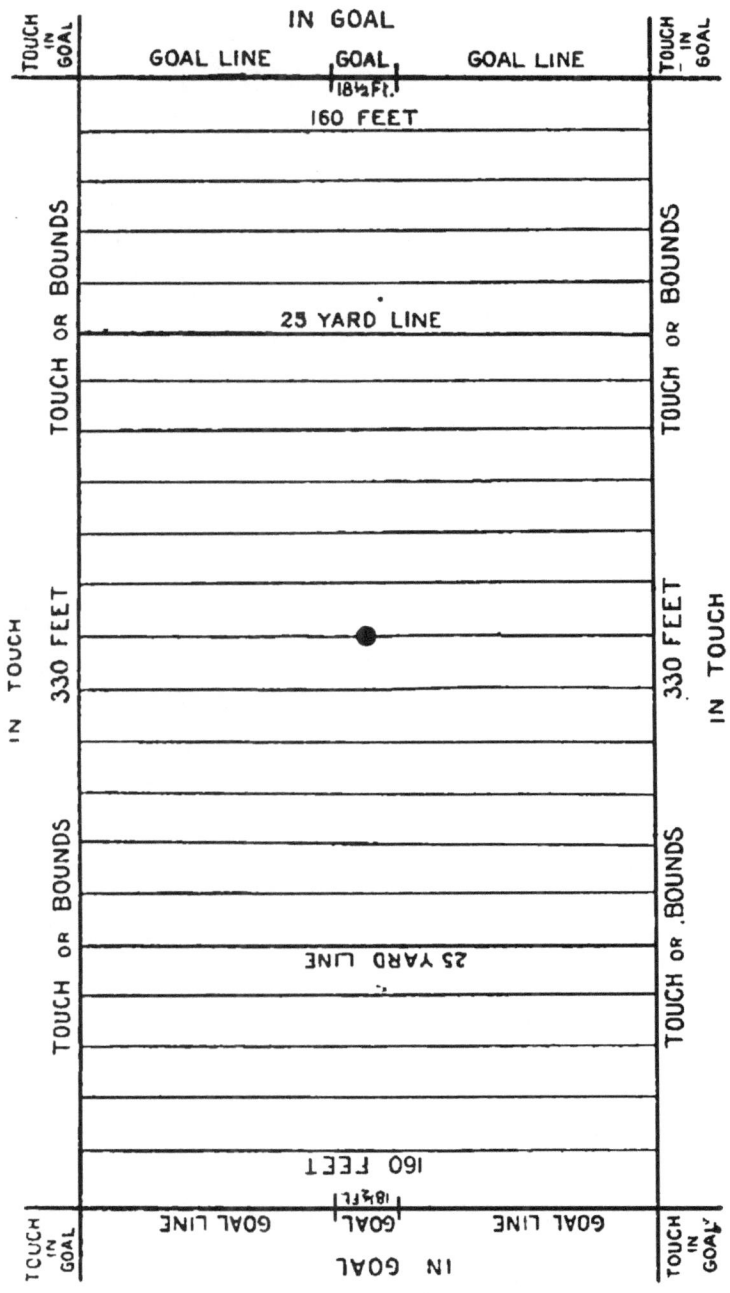

IN GOAL

GOAL LINE   GOAL   GOAL LINE

|18½Ft.|

160 FEET

TOUCH IN GOAL

TOUCH OR BOUNDS

25 YARD LINE

TOUCH IN GOAL

TOUCH OR BOUNDS

IN TOUCH

330 FEET

330 FEET

IN TOUCH

TOUCH OR BOUNDS

25 YARD LINE

TOUCH OR BOUNDS

160 FEET

|18½Ft.|

TOUCH IN GOAL

GOAL LINE   GOAL   GOAL LINE

TOUCH IN GOAL

IN GOAL

INTRODUCTORY CHAPTER

FOR

BEGINNERS AND SPECTATORS.

American football is played on a rectangular field, three hundred and thirty feet long and one hundred and sixty feet wide, enclosed by heavy white lines marked in lime upon the ground. For the convenience of the referee in fulfilling his duties, the field is marked by additional lines five yards apart crossing from side to side, the fifth from either end being indicated by an especially heavy one known as the "twenty-five yard line." The "center of the field" is located at the middle point of the eleventh line.

At the center of the goal lines at each end of the field two goal posts, from fifteen to twenty feet high, are erected eighteen and one-half feet apart, and connected by a cross-bar ten feet from the ground. Two "teams" of eleven men each contest in the game. Seven of them, called the rushers, or forwards, stand opposing a corresponding seven of the opposite eleven, whenever the ball is down for a "scrimmage." The one in the middle is known as the center rusher, or center, and on either side of him are the right and left guards, the right and left tackles, and the right and left ends, respectively. The four remaining players are the quarter-back, right and left half-backs, and the full-back, who stand behind

(7)

the line of rushers and occupy positions which vary ac·
cording to whether they or their opponents have the ball.
The positions which the players will occupy when about
to execute the different movements of the game are shown
by the diagrams in the chapter describing the various
evolutions. At the beginning of the game the ball is
placed at the center of the field. The side in possession
of the ball constitutes the side of attack, and endeavors
to carry it down the field by kicking or running with it,
in order to place it on the ground behind the opponents'
goal line. The other side, forced to act upon the defensive, are drawn up in opposition, and strive to check their
advance and to get possession of the ball themselves, so
that they may no longer act upon the defensive, but
become, in turn, the attacking party.

The rules of the game (see final chapter), place certain
restrictions upon the attacking side and upon the defense,
and it is the attempt made in accordance with these rules
by each side to retain the ball in their possession and
carry it down the field through all opposition, in order to
place it behind their opponents' goal, which furnishes in
rough outline the essential features of the American
game of football.

Before the game is to begin the captains of the respective teams decide by a toss of the coin which side
shall first be given possession of the ball. The side having
the ball then places it down upon the center of the field
and arrange themselves in any formation which they desire, behind the line on which the ball is placed, in preparation to force it into the enemy's territory. The side
acting on the defensive are obliged to withdraw ten yards
toward their own goal, and are there drawn up in opposition to await the attack of their opponents until after
the ball is put in play.

As the "center rusher" of the attacking side puts the ball in play by touching it with his foot and passing it back to some other player for a run, or a kick down the field, the rushers upon the defensive side are at liberty to charge forward to meet the attack. The clash following this charge constitutes the first actual encounter of the game.

When the runner with the ball is caught, or "tackled," thrown upon the ground, and there held so that he can advance no further, he calls "down," whereupon the ball is "dead" for the moment, and cannot be carried forward or kicked until the center rusher again puts it in play according to rule.

As soon as "down" is called, an imaginary line, crossing the field from side to side and passing through the center of the ball, immediately comes into existence. Each player must remain on the side of this line toward his own goal until after the ball is "put in play," and it is one of the duties of the umpire rigidly to enforce this regulation. Should any player cross this line and fail to return before the ball is "snapped back" it constitutes an "off side play," for which the rules provide a penalty.

To again put the ball in play the center rusher places his hand upon it at the spot where "down" was called. The rushers then "line up" opposing one another, the line of attack being drawn closely together for a greater concentration of energy, while the defensive rushers are slightly spread apart to facilitate breaking through the line and stopping the advance, when the ball shall be put in play. The captain upon the attacking side then shouts some signal, understood only by his own men which indicates the evolution that he wishes his eleven to execute ; whereupon the center rusher puts the ball in play by "snapping it back," that is, by rolling it back between his legs.

Immediately behind the center rusher the quarter-back has taken his stand. He receives the ball as it is "snapped back" and instantly passes it to one of the half-backs or a man in the line, for a run, or to the full-back for a kick down the field. Thereupon the first "scrimmage" of the game takes place as the opposing team attempts to break through the line and stop the play.

One side is not allowed to retain indefinite possession of the ball without making gain or loss. The rules provide that if the side having possession of the ball shall fail to make an aggregate gain of five yards, or a loss of twenty yards, in three consecutive "scrimmages", the ball shall be forfeited to the other side at the spot where it was last down.

Advances by running are made by the player directing his course through one of the six openings in the rush line, or around the ends, according as the signal may direct. The signal also indicates the player who is to receive the ball. The runner is assisted in his course by the players who border on the opening through which he is to go. These seek to enlarge the space by pushing their opponents to one side. He is further assisted by others of his own players, some of whom precede, to "block off" the opponents from "tackling" him in front, while still others follow to push him further if he is checked. The players who are to precede and the players who are to follow change with the play according as each man is enabled by his position to adjust himself to it.

Four points are scored when one side carries the ball across the goal line and makes a "touch down." The side making the "touch down" is then allowed to carry the ball out into the field as far as they may desire in a line perpendicular to the goal line and passing through the point where it was "touched down," in order that one of their number may attempt to kick it between the goal posts above the cross-bar. The other side meanwhile are obliged to take their positions behind the goal line. Should the attempt be successful, it will constitute a "goal," and two additional points be added to the score. But whether the attempt be successful or not, the ball must be delivered to the other side, who will take it to the center of the field and put it in play in the same manner as at the beginning of the game.

If the ball can be kicked between the posts and above the cross-bar by a "drop-kick" or "place-kick" by any one of the players, without having been previously carried across the goal line, it will constitute a "goal from the field," and will count five points.

In case the ball is kicked or carried across the boundary line on either side it will be "out of bounds" and must be brought into the field at right angles to the line at the point where it crossed. This is done by the side which first secures it after it passes out of bounds.

It is usual to bring the ball into the field from ten to fifteen yards and then to place it upon the ground for a "scrimmage" as from a regular down ; though the ball may be passed to any one of the players, in at the point where it went out, provided that it is thrown in at right angles to the side line ; or it may be "touched in" at the same point.

The game is divided into two halves of three-quarters of an hour each, and the team succeeding in scoring the greatest number of points during that time are declared the winners.

The reader should thoroughly acquaint himself with the rules in detail, before passing on to a study of the book.

TRAINING.

In the early days of college athletics and amateur sports the popular belief was universally accepted that a most rigorous diet must be entered upon if the young aspirant for college honors would fit himself properly to represent his alma mater in the boat, on the running track, or in individual contests. Many an alumnus who pulled an oar on the crew in the fifties and sixties, will recall visions of raw beef, a limited bill of fare, and a prescribed daily amount of water that made the training of thirty years ago a hardship for which dim dreams of possible glory seemed a doubtful compensation.

These old ideas have now changed almost entirely, and the young collegian of to-day, who secures a position on any one of the college teams, and obtains a seat at the "training table," is an object of envy rather than of compassion to his classmates. The training table diet of to-day is almost sumptuous, and few men in college enjoy better living than the members of the university athletic organizations. Roast beef, lamb chops, beef steak, roast lamb, and broiled chicken, oatmeal, rice, mush, and the cereals, potatoes served in all styles but fried, stale bread, onions, garden vegetables in season, eggs, dry toast, apple sauce, baked apples, prunes, grapes, oranges, figs, dates, and fruits in season (with the exception of raw apples), rice and bread puddings, furnish an abundant variety from which to choose.

A few things only are put upon the proscribed list: Pies, cakes, salads, all forms of pork, veal, rich dressings, fried food, ice-cream, confectionery, soda water, so-called soft drinks, (and it is needless to say drinks of a stronger nature,) tea, coffee, and chocolate, should be cheerfully and absolutely given up. From the first day of training it should be rigidly enforced that all pipes, cigars, and cigarettes be laid aside, absolutely, until the contests are over.

Regularity in all the daily habits of life is of the greatest importance. The hours for rising, for meals, and for retiring should not vary from day to day; and in so far as it is practicable to do so, it would be advantageous to have the regular practice come at that portion of the day in which the important games of the season will take place.

That the football player should have long hours of restful sleep is a point too frequently overlooked. While it is impossible to state a definite time that shall apply to all cases, a sleep from ten o'clock in the evening until seven the next morning, and a short walk before an early breakfast, will be found to be of the greatest benefit in all instances. Probably a large proportion of the cases of over-training, that occur during the football season, are caused by late hours, irregularity of habits, and insufficient rest. Had these points been carefully attended to, the hard work upon the field would have produced no hurtful result. When the recreation period of the players makes it necessary that the daily practice shall come immediately after the noon meal, it will be found more healthful to have the practice hour preceded by a light lunch, and postpone the hearty dinner until night. But should the daily play come in the morning, or in the middle of the afternoon, it will be better to have the dinner hour at noon.

Over-training is something which is much easier to

prevent than to remedy when once it is an accomplished fact. In preparatory schools, where a less violent and less tiring system of training is followed, no thought need be given to this point, but in the larger colleges one or more cases of over-training among the valuable men is apt to occur toward the end of a season of hard work.

Should any one of the players get into this condition, he should be given an absolute rest for several days, and then be allowed to play only part of the time during each remaining day of practice. An immediate change of diet with a removal of all training-table restrictions, will also be found of value.

When a faithful worker finds himself coming upon the field day after day with a worn and tired feeling, no longer able to play with his former dash and energy, and his speed gradually decreasing, he should at once suspect that his muscles are becoming over-tired, and so fatigued that they cannot recuperate between one day's work and the next.

The practice of drinking water during the game is exceedingly bad, and never should be permitted, though rinsing the mouth is admissible. The best results will be obtained if no water whatever is swallowed until more than an hour after the practice is over. The habit which some players have of chewing gum during the game is pernicious. After the first week or two has passed, the mouth will be found to be far less dry where no gum is used, than where a constant flow of saliva is kept up by the act of mastication.

During the season there undoubtedly will be a number of rainy days. These by no means should be lost. As a rule, it is best to practice upon the field as usual, since the most important game of the season may come in bad weather, and the experience of having frequently played in the mud with a wet and slippery ball will prove invaluable.

On special occasions light work in the gymnasium, tackling the bag, and practicing the signals indoors, may be substituted with advantage. Every team should be provided with a tackling-bag. This may be made of leather or canvas, and should be from four to five feet long, a foot in diameter and stuffed with hay, hair, or excelsior, to represent the body of a man. No better practice can be had for low hard tackling than to have such a bag suspended by a long rope from a rafter in the gymnasium over a number of floor mats, letting the men run half the length of the floor and spring for it from some ten feet away as it swings slowly backward and forward. But except on such special occasions when no out-door practice is taken for the day, indoor gymnasium work should be given up, as the exercise upon the field demands every energy.

During the last few weeks of the season, when the final eleven has practically been decided upon, and team play is being developed, an opportunity should be found each day to send the eleven up and down the field in their regular positions, upon short runs of from five to fifteen yards, with no opposing rush line drawn up against them, in order that the signals may be thoroughly drilled into each player and substitute, and all learn to work together as one man. It is of the highest importance to have a number of substitutes, each of whom is thoroughly acquainted with the signals, as the replacing of a player in case of accident by one in the slightest degree unfamiliar with the signals will destroy team play and cause the side a loss much greater than the value of the man who has left the field.

The number of regular games a week a team can play

2

to advantage cannot be definitely stated. The condition of the men and their especial needs must determine this. As a rule, more than two match games a week cannot be played if the best results are to be obtained. A hard game should not be played within less than a week before one which is considered to be of great importance, if it can be avoided, on account of the danger of having a valuable man disabled, and in order that there may be an abundance of stored-up energy upon the day of the important contest.

During the last few days before the final game, the practice should be short, but sharp while it lasts, with a considerable amount of time devoted to practicing the signals, falling upon the ball, and perfecting team play. On the day immediately preceding the game an absolute rest should be taken.

It is a mistake to attempt to play the full hour and a half on each day of practice throughout the season. About two half hours of sharp work, with a rest of five minutes between, will produce the best results, and in the earlier regular games each half should be limited to thirty minutes.

The daily practice of the team upon the field will not afford sufficient opportunity to the backs to become proficient in kicking and catching the ball. When it is possible, a half hour should be devoted by them at some other portion of each day throughout the entire season to punting, catching, and goal kicking. Numerous minor sprains and bruises will necessarily be received during the season, for which hot water and flannel bandages will be the best remedy.

In case of a sprained ankle or a serious bruise to one of the muscles of the leg, a long period of disability may result from continued playing, and the captain should insist that a player so hurt should leave the field at once. A thin leather anklet had better be worn inside the shoe by each player in the team as a safeguard and protection.

When a man has a bruised and sensitive knee, a moistened sponge, the size of a fist, placed just under the knee cap will afford relief and protection. Sprains and bruises of a serious nature are more liable to occur during the first few weeks of practice than at any other time in the season. This is due to the fact that many of the men have just returned from long vacations of ease and idleness, and their muscles are not ready to endure the sudden strains and wrenches to which they immediately find themselves subjected. The careful captain will see to it that the promising new candidates for his team and the old men are all gathered together from one to two weeks before the season of actual playing is to begin, and put through a series of light exercises, given short runs, made to pass, kick, and fall on the ball, and are given such general light work for wind and muscle as shall enable them to engage in the regular practice without danger. Thick sweaters and overcoats should always be in readiness to put on after playing, and proper care taken to guard against catching cold.

Cleanliness is a hygienic necessity during the football season, and every team should, if possible, have hot and cold water shower baths connected with their dressing rooms.

Long hot baths are weakening, and should be avoided; though upon special occasions, when a cold has settled in the muscles, a Turkish bath may prove of great value.

The captain's word upon the field is absolute law, and should be followed with unquestioning obedience.

THE CENTER-RUSHER.

The prevailing idea·in time past has been that the largest and heaviest man who could be procured should be used for the center-rusher, or snapback of the eleven. So universal has this idea become that it has long been a common joke to say of an especially large and stout person: " He would make a good center-rusher." Every new team formed, as a rule, selects the center according to this axiomatic fallacy. It is easy to see how this principle of selection became established under the old pushing style of game, and it still should hold sway, provided it brings with the selection certain qualities of mind, and certain physical capacities, which will enable the center to be one of the most active and effective agents on the field.

The center occupies a unique position on the eleven in that he starts the play after each down, and is the only member of the team who cannot run with the ball from a scrimmage, because it is impossible to make him a third man advantageously. His work, therefore, is limited in that particular. By reason, also, of his having to protect the quarter-back after he snaps the ball, and because he is invariably entangled with the opponents, it is impossible for him to become a valuable running interferer. What work in interference he is able to do is limited to blocking the opponents from breaking through the line, or running behind their own line to head off the runner with the ball at one side. Possibly, when very clever and swift, he may be able to cut across the field to interfere with a half-back or the full-back. The center should

(18)

make a practice of doing this latter work on every play around the end, and on every play between the tackle and end. Perhaps he may not be able to get ahead of the runner, but he will be of valuable assistance by checking some of the opponents from running behind their line and tackling him. Now and then, also, he will be able to get ahead of the runner and go down the field with him.

From these statements it might appear that it did not matter especially whether the center rusher was a slow runner or not, and that emphasis should be laid on his possessing size and weight, which are understood as necessary to the proper filling of that position. The truth is, that while a slow runner, if he has cleverness for that position and is strong and weighty, will be able to do fairly well as a center, he cannot begin to be as serviceable to his team as if he were also a fast runner. Granting that a fast runner will not be able to do much interfering, or running with the ball, he will still be able to use his speed most helpfully in breaking through the line to tackle; in crossing over to one side to head off a runner; or in going down the field on a kick. Furthermore, his speed will be most helpful in playing a quick game, because he is thus able to follow the ball so closely that there will be no delay in putting it in play. This is a most important point in the center's play. He must be on hand to receive the ball the instant it is down.

It is impossible to play a quick game where the center lags, or to prevent one on the part of the opponents. When there are not many large men who are fast runners it is better, perhaps, to place the speedy man in the position of guard and take a slower man for center.

The ideal center will be one who is swift of foot in addition to his other powers. He should be a large man, not a ponderous man, unless he is quick and strong. He should be especially strong in his legs and back, for he must stand steadily on his feet against the continuous pushing and wrestling which he receives, directly from the opponents, and incidentally from the guards on either side of him. If he is easily moved, or toppled over, he will be likely now and then to snap the ball poorly, thus making the quarter-back uneasy and flurried in handling it. Steadiness is a most necessary part of the center's work and it cannot well be overlooked in the selection of a man to fill that position. Further, as in every position on the eleven endurance is a prime requisite, so is it in this. More of it is needed, however, than in most others, because the work is much harder. No short-winded, fat man can long stand the hard work of that position, if he does his duty. Not only is great physical labor required of the center, but he must also be constantly subjected to knocks and bruises from the plunging and tearing of the rushers and half-backs as they try to break through the line.

No man, therefore, can play in this position who is not physically courageous, and who is not able to rise to his work after each assault with new grit and determination. He should be a man who is cool and collected at all times; combative, but never losing control of his temper; one who endures worrying without being rattled by it; one who never gives up and is bound to conquer. Nowhere in the line is there need for such steadiness as in the center. From him every play starts, in a scrimmage, and a little unsteadiness on his part will be likely to make havoc with the quarter-back's work, and hence with the offensive play of the whole team. Nothing can be more fatal to quick and steady play, for it is sure to produce hesitancy in action in some of the players, with hurried action in others.

In assuming his position for a scrimmage, the center may follow either of two methods of standing, when snapping the ball : one, where one foot is placed back for a brace, the ball being snapped between the legs and a little to one side ; the other, where both feet are widely spread to interfere with opponents, as they attempt to break through, and to avoid getting into the way of the ball which can be snapped straight back. Where the first position is followed, the center should be able to work equally well with either foot forward, in order to secure certain advantages in handling his opponent. The center-rusher should make a study of the best way of snapping the ball back, and then hold it the same way every time. He should confer with the quarter-back on this point, as the latter is to handle the ball, and it may be easier to take it when snapped in a particular way.

There are two methods followed in snapping the ball : one, in which the ball is held on the small end and sent back swiftly, with little effort, in such a way that the quarter-back catches it in the air all ready to pass ; the other, where the ball is laid on its side and rolled along the ground to the point where it is stopped by the quarter-back and then picked up in very good position for passing. This latter method is more generally used because it does not require as delicate work on the part of the center in giving the snap ; but speed is sacrificed by it and there is greater liability that the ball shall be deflected from its course by touching the legs. It would be well for the center to learn to use either hand in snapping, for it will often prove an advantage. The center-rusher will do well to make a study of snapping the ball by both methods of standing, and by both ways of holding it until he settles on the one best suited to him. He should then practice this against an opponent until he is able to stand firmly on his feet and send the ball back accurately, at a uniform rate of speed each time. In case the ball is placed on end, it is better to have it lean toward the opposing center at an angle of about sixty degrees. It can be held more firmly in this position and can also be sent back more swiftly, with a bound into the air. Care must be taken not to send the ball *too* swiftly. While the center is practising to secure steadiness, accuracy, and uniformity in snapping the ball, he should likewise practise getting his opponent out of the way.

In putting the ball in play, the center has the advantage of being able to select the time to snap and he can choose it to meet his own purpose. Besides, he knows the exact instant when he intends to send the ball back and can get the start of his opponent. The center, therefore, is master of the situation when he has the ball. It is for these reasons that he can frequently be down the field on a kick as soon as the ends, and yet not expose the full-back to great danger in having the ball stopped.

There are various ways for the center to handle his man and get him out of his way. He may plunge forward at the instant he snaps the ball, carrying his opponent before him ; he may lift him to one side or the other, according to the play called for and the position of the opponent ; he may fall on him if he is down too low ; or he may get under him and lift him in the air, if his opponent reaches over him.

In any one of these methods, the opportune moment must be seized like a flash and the action be quick and powerful. A slow, strong movement will never succeed. Long and faithful practice is necessary before the center can acquire this quickness and power. In his eagerness to take advantage of his opponent, he must never fail to wait for the quarter-back's signal before snapping the ball. A little forgetfulness on this point might prove disastrous.

The center can be a most valuable man in defensive play if he understands his position. By giving his opponents a quick pull forward or to one side at the instant the latter snaps the ball; by lifting him suddenly backward; or by grasping his arm, the center can frequently break through more quickly than either guard or tackle. Whenever he succeeds in getting through, he will be a strong obstacle to all dashes between himself and the guards, and he will sometimes be able to interfere with the quarter-back's pass. Another way in which the center may play on the defense is to spend all his energy for a moment in getting his opponent out of his way and then spring at the runner. In this case the center must throw off his opponent quickly, and not allow himself to be carried backward. At the same time he must not attempt to break through the line.

When the play is around the end, or even at the tackle, the center should move quickly from his position and pass around behind his own line to meet and tackle the runner. When the opposite side is about to kick, the center should do his utmost to break through the line and stop it; but sometimes it may be better instead to make an opening for the quarter-back. He is helped in doing this, by the opposite center himself, as he plunges forward to block him. In such a case a good opening can be made for the quarter-back, if the center will place himself in front of his opponent a little to one side, and then pull the latter forward to the right or left. The guard at the side on which the opening is made should know of this plan so that he may not spoil it, either by pushing his opponent in the path or by getting in the way himself. If there is danger of his doing this, it will be better for him to help enlarge the opening for the quarter-back.

On the defensive the center may play a little to one side or the other of his opponent, or directly in front, to suit the situation. It is most unwise for the center to assume the same position every time, for by so doing he gives the opposite center only one problem to work out and that one probably the same each time. Where the center takes an extreme side position, unless he does it just before the ball is snapped, he gives the captain of the other eleven a fine chance to call for a play which will take advantage of the situation.

There is abundant opportunity for the display of head-work in outwitting the opposing center in breaking through the line. The line is so compact at this point that it is not an easy task to slip by, especially as the opposing center is watching to take his man at a disadvantage Various methods are resorted to in breaking through the line. Sometimes the center, acting on the defense, is thrown head foremost to the ground by a quick, hard pull, the attacking center stepping aside or over him as he falls. He may also be turned sidewise just enough to slip past him, or he may be lifted back perhaps into the face of the runner. The most common method employed by the center in getting through is to catch the arm of the opponent on the side on which it is desired to go through, give it a jerk, and dash into the opening.

The center in defense must insist on the ball being down where it belongs. Some center-rushers have a way of moving the ball forward several inches further than it should be. There is no occasion for generosity under such circumstances, and the center must feel that it is his duty to stand up for the rights of his team by constantly guarding against any infringement of this kind. On the other hand, a constant bickering over an inch or two of ground may be made of such importance that the game is interfered with and delayed to such an extent that a much greater gain would have resulted were the ball put in play the instant the signal called for it.

A good referee will see to it that the ball is snapped each time from the proper spot.

It is always the duty of the center-rusher to keep close to the opponent who brings the ball in from the side line, in order to protect the rights of his team. Likewise, it is well to " pace in " the opponent who brings the ball to the twenty-five yard line, in order to prevent a quick play being made when his own side are not in position. The guards assist him in this.

THE GUARD.

The main work of the guards may be summed up as blocking, that is, guarding; making openings for the passage of the runner whenever certain signals are given; running behind the line to interfere for the man with the ball; running with the ball occasionally; breaking through the opposing line to interfere with the quarter-back in passing the ball; and tackling the runner or stopping a kick. The guards and the center have the most laborious work on the eleven, if they do their duty, for they practically have no respite from hard work. They must bear the brunt of the heavy plunging of their opponents through the center, and at the same time struggle to break through the opposing line, which is doing its utmost to prevent them. They must do this without a let-up just as long as the other side has the ball, and, moreover, in that part of the line which is most compact. Then, when their own side has the ball, they are expected to use their strength and wits from the moment the ball is put in play until it is again down, in blocking, making openings, and in interfering for the player who is attempting to run. Further, they have little time to catch their wind, for almost the first point which should be drummed into them by the captain or coach is to be always on hand the moment the ball is down, to make or prevent a quick play. It can be truly said that no team is well trained until the center part of tho eleven, as indeed the whole team, is prompt on this point. While the guards have all this hard work, they seldom have a chance to distinguish themselves, either by a run, a clean tackle, or a fine interference which is apparent to the untrained eye of the spectator. On the other hand, it does not take much yielding at the center to bring forth the criticism that that part of the line is weak.

On account of the nature of their work, the guards should be large and powerful, like the center. It is even more necessary that they should be quick, agile, and swift, than the center, because the guards should always go through the line when the opponents have the ball. On their success in doing this largely depends the strength or weakness of the team's defense.

The chief point in defensive play is to tackle the runner before he reaches the line, and the guards are large factors in doing this. Unless this is done, the ball can be steadily carried down the field when not lost by a fumble, for any team is able to gain five yards in three consecutive trials when the runner is allowed to reach the line each time before being tackled. Any means, therefore, which the guards can employ to interfere with the quarter-back before he has passed the ball, or the runner before he has reached the line, should certainly be used. All the strategy and tricks known in wrestling which can be applied to the situation should be eagerly sought and practiced. The great point to remember is to apply the power quickly and hard, to summon all the strength for the initial effort, and to work desperately until free from interference. Only by doing this can the guards hope to break through and secure the quarter-back or runner behind the line. Slow pushing, however powerful, will accomplish little. If held in check until the runner and the pushers strike the line it is only a question of how many yards the runner will gain before the mass breaks and falls forward.

In applying his power the guard, as well as his companion rushers, has an immense advantage in being permitted to use his hands and arms freely in getting his opponent out of the way. This enables him to put into practice all the skill he possesses in handling an opponent who is allowed to block only with the body. The guard also has another advantage in being free to move whenever he pleases, but he must remember that the opening for the runner may be made on either side of him and be careful not to give his opponent help in making it. It assists the guard greatly in breaking through if the tackle draws out the opposing line as much as is wise in a good defense. This separation should be wide enough to allow the players in defense to break through easily without interfering with each other. It is also usually helpful in breaking through to be restless, but cautious at the same time, in order not to give the opponent an advantage.

The guards and the tackles especially should watch for signs which shall indicate what the play will be, and then go through the line as low as possible for a tackle. They should break through to the right or left of their opponents as seems best at the moment. In order to break through quickly they must have their eyes on the ball when it is snapped and spring forward the instant it is put in play. Quick glances may be cast at the opponents while still constantly watching the ball.

The guards, with the center, are usually called upon to meet the heavy charges in the opening plays from the center of the field. These, as a rule, come in the form of wedges. Two points should be carefully regarded by these center men in attacking a wedge: first, to approach the wedge with the body bent in a position for greatest power and for meeting the wedge low down; second, to focus on the mass in such a way that it cannot break through between them without being separated, and so giving the guards a chance to tackle the runner. In doing this it should be the aim to focus as nearly as possible upon the point of the wedge, in order to check its advance and throw the forwards back on the runner. The runner will then be forced to come out, if he has not already become entangled in the mass. In making the attack the guards and center should run with dash and determination, at the same time watching closely for the runner and trying hard to tackle him.

Two successful ways of attacking a wedge have been originated. One member of the center trio will sometimes jump over the heads of the forwards and try to fall on the runner and thus secure him, or he will hurl himself headlong at the feet of the oncoming wedge and cause it to trip over him. To make either one of these attacks well the player must be perfectly fearless, and should also use good judgment. In the former case the player must time his jump and not land short of the runner, or he will be pushed quickly to the ground or carried along on the heads of the forwards; neither must he jump so far over that he will miss his man. If he throws himself in front of the wedge he should not do it too soon, lest the wedge will be able to avoid or step over him.

When a wedge is formed in the line on a scrimmage the guards and center must be sure to get low, or they will be carried along before it. The *point* of the wedge must be held in check. In resisting the attack of a revolving wedge the guards should separate slightly from the center and join with the tackle in trying to penetrate the mass to secure the runner. This should be done in such a way that the defense shall not be weakened. Care should also be taken by the side of the line away from which the wedge revolves not to add impetus to it by pushing too far.

The position of the guard varies slightly in defense and offense. In offense the first thought must be to protect the quarter-back until he has passed the ball; his next to block his man long enough to prevent him from reaching the runner. His third thought, which may also influence the way he stands while he attends to the former work, is to make the opening if the play is in his quarter. His fourth thought, which will be influenced by his first and second, is to get in his interference ahead of the runner when practicable, or follow him as closely as possible and do what he can to assist. In fulfilling all these duties he will be limited in his freedom of movement. He cannot stand too far from the center rusher, and he may be compelled to stand shoulder to shoulder with him.

Further, he will have to assume a position which best enables him to carry out his duties. It may be well for him to stand with both feet on a line, or it may be better to have one or the other foot behind, according to his purpose. It is nearly always better for him to bend forward, or even to get down very low if his opponent tries to get under him. The bent-over position is better for meeting attacks, because the weight is well forward and low down and the body is better braced and not so much exposed to effective handling. In this position, also, one can move forward better for making an opening.

In blocking the legs should usually be spread widely apart. They should not be spread so much, however, that the guard will not be able to move quickly whenever his opponent shifts his position. In blocking, as in breaking through the line, the guard should try hard to get his power into action before his opponent. This can be best done by a shoulder check.

The general position of the guard must be determined by the play in hand and the way the opponent stands. He may be forced to move out a little because his opponent does so, but he must be careful that the opening between him and the center is not occupied by the quarter-back or some other free player, in which case the tackle will sometimes be obliged to step in and take the opposing guard. Neither the guard nor any other rusher except the center should ever take a fixed position in standing.

On the defensive much depends on strong blocking by the guards, for weak blocking is fatal at the center of the line. The quarter-back, being so near to the guards, is in imminent danger in case of weak blocking, and he can little afford the loss of a fraction of a second in handling the ball, much less a fumble. Under these circumstances, if a fumble occurs, the quarter-back must always fall on the ball and not run any risks of losing it. Furthermore, in weak blocking the runner has little chance on a dash into the line, for in place of an opening he finds an opponent. "Block hard" has come to be one of the axioms of the game. Blocking for a kick is treated fully in the chapter on team play.

The guard has an advantage over the center in making an opening for the runner in only one particular, and that is that he is freer to move in his position. The center rusher is largely dependent on the position which his opponent takes in standing to help him out in this matter, since he cannot move his relative position from the opposing center more than the latter allows; but he can often influence that position to suit his own purpose. By clever generalship and strategy he may be able to induce his opponent to do the very thing he needs to help him out in his play. Some of the ways of handling an opponent are given in the description of the duties of the center rusher.

When the guard is going to run with the ball he should take a position which will enable him to get away from his opponent quickly, but he should not make his intentions evident. For this reason it is better for the guard, as well as for the tackle, not to take a set position until the signal is given; but if one is taken, let it be such that it would not make it necessary to change in order to run with the ball. The one who is to run with the ball should seek in every way to conceal the purpose of the play.

The guard is in the most difficult position from which to get under headway in order to run with the ball. As commonly played, the guard swings round the quarterback and dives into an opening between the tackle and guard on the other side of the center. The very beginning of his run is the most difficult part. He cannot run fast from his position, for he has only a step or two to make before he must turn sharply around the quarterback and run in almost an opposite direction. If he runs back too far he will be tackled before he reaches the line, and if he turns in closely, he is likely to run against his own men as they are struggling with their opponents. It needs, therefore, careful judgment and a great deal of practice to be able to run well from this position.

Long-legged guards, as a rule, find it easier to take a long step backward with the foot next the center, and use that as a purchase from which to circle around the quarterback. Some guards prefer to take three or four short, quick steps in making the turn around the quarter-back. Any way which will enable the guard to get under headway most quickly is the method which should be used. It will be easy for the quarter-back to place the ball in the guard's hands, and it will probably be better for him to carry it under the arm away from the center.

When the guard runs around to interfere, he should place himself so that he can get away quickly and not "give the play away." If the guard is to run around in order to interfere by getting ahead of the runners, the quickest possible start is necessary. There must be no delay whatever, even when the guard is a fast runner, or else the runner with the ball will have to slow up so much that he cannot make the play. Whenever the guard runs around to interfere or to run with the ball, the tackle should keep the opposing guard from following him. The guard can sometimes do this himself by pushing his opponent back just as he starts, but it must be done in such a way that it will not delay him.

THE TACKLE. ·

The tackle occupies the most important position on the rush line. It is possible to get along with a lumbering center and slow guards if they are able to block well and make good openings, but it is not possible to have slow tackles and play good football at the same time. The position which the tackle occupies in the line explains this, and it is best appreciated when it is understood that the tackles should take part in more than half the defensive work of the team.

The tackle occupies the most responsible position because he assists in checking two distinctly different styles of play. On the side toward the center he is to help the guard in blocking the heavy plunges which are frequently aimed at that point of the line, while on the other side he has to work with the end-rusher against all plays between them and on all plays around the end. To play this position properly on the defensive, therefore, requires a master mind and an equipment of physical capacity and skill unequaled by any position on the eleven.

Next to the half-back the tackle, from his position in the line, has the best opportunity for running with the ball. In fact, he can be used with telling effect, if a good runner, in supplementing and resting the half-backs. Again, he is the end-rusher's chief assistant in going down the field on all kicks, and he must be under the ball almost as soon as the end himself, in order to prevent the catcher from dodging inside the end men.

(34

The points mentioned are sufficient to show that the tackle should be a man of considerable weight, because he has to bear a great deal of the heavy plunging into the line. The greater the weight the better, provided, of course, that the other requirements are met. As a rule, it is rare that a man weighing over one hundred and eighty pounds can meet these requirements, and it is more often that men weighing one hundred and sixty-five or seventy pounds are selected for this position on the best teams. The general build of the man also qualifies his usefulness. The one hundred and sixty-five pounds will be much more effective in a man from five feet six to five feet ten inches in height than in one above that height. In truth, the man of stocky build can usually fill this position much better, because his weight is nearer the ground and he is always in a position to make a low tackle. As a great deal of his tackling should be dashing and brilliant, right in the midst of interference where he must throw himself instantly, a tall man would be at a disadvantage. A thick-set, round-bodied man with large arms and legs would also be a much harder man to stop when running with the ball.

Of equal importance with weight, the points which should determine the selection of the tackle are agility, speed, and the ability to tackle in the face of interference. The name of the position indicates the work of the player. He is to *tackle*. Even speed can to a small degree be dispensed with if the man is quick and agile and is a sure tackler. Quickness in getting through the line, agility in avoiding interference, sure tackling, getting down the field on a kick, and running with the ball are essential qualifications to look for in selecting a man to fill the position of tackle.

The tackle must be endowed with more than the ordinary amount of shrewdness and judgment. To a certain extent this can be acquired by long practice, but the tackle must be of quick perception and good judgment naturally in order to play the position in the best manner.

When acting on the defensive the distance which he should stand from the guard and the manner of going through the line, either to the inside or outside of his opponent, should be determined by previous judgment as to where the play is to be made and influenced by an instantaneous perception as the play starts. The position, too, must be taken with the utmost caution and selected at just the right distance from the guard to best meet the play and still be able to defend his position on either side. There is need of the closest and quickest observation and cleverest judgment.

Moreover, as many of the plays cannot be determined beforehand, such a position must be taken as will best enable the tackle to check any play which can be made. He must then be on the alert for the very first indications of the play and act on them, and at the same time he must still keep the closest watch for later developments which change the direction in which the ball will finally be carried.

Playing up close to the guard is always dangerous unless it is necessary to do so in order to stop a wedge play, for the tackle could then be blocked in very easily from helping, if an attack were made on the space between himself and the end man, or in a play around the end. He therefore would cut himself off from defending two-thirds of his territory and the most defenseless part of the line. Playing far away from the guard is also dangerous, for he then leaves the part of his territory which is nearest the opposing half-backs too much exposed and gives his opponent a chance to block him off from defending it. Of course, if the tackle were free from the checking of an opponent, he could play some distance away from the guard and still defend the space between them ; but the fact that there is a player opposite who is giving all his attention, wit, and energy to securing an advantage over him, gives a turn to the problem which he cannot ignore in making his calculations. The tackle takes a certain position ; the opponent takes one also. It may be a little to the right or a little to the left of him, or it may be directly in front of him. The tackle may change his position a little and then the opponent perhaps change his, but their relative positions may, or may not, be changed ; or possibly his opponent may remain in the same place. Just this action or inaction on the part of the opposing tackle is sufficient to help him determine how he should play in his defense, and is one of the signs to be considered in deciding upon his own position and action.

The tackle should usually play right up to the line, on the defense. Sometimes with a very quick opponent, it may be better to play a little back from the line. He should be restless, and on the alert for an opportunity to go through on the side of his opponent offering the best advantage. He should watch the ball closely and spring the instant it is snapped. His course of action in reference to his opponent must be to get him out of the way as quickly as possible. It may often be best for the tackle simply to drive his opponent back with hard, quick pushes. This might frequently be best when the play is between him and the guard, because the time for preparation to tackle is exceedingly short before the runner will be going past, and the whole attention must be given to securing a momentary freedom from interference, for a quick spring. The tackle has a great deal of this quick tackling to do because the runs are so frequently made in his region. Much of this also must be done right in the midst of interference, when the only chance to get the runner is by hurling himself headlong at him as he passes.

On end plays the tackle must break away from his opponent as quickly as possible. He will have no time then to carry his man before him except, perhaps for an instant, as he pushes him back to get by him. Yet he must make sure to knock his opponent sufficiently off his balance to prevent his following him and giving him a shove at a critical moment. In defense on an end play, everything depends on the tackle reaching the runner before he begins to turn in order to circle the end, and before he has swung in closely behind his interference. The runner then has not yet gotten under full speed and the interferers are somewhat scattered and looking toward the end. The tackle has the best chance for defeating end runs ; in this he is ably seconded by the end man, the two working together, in fine team play.

The tackle must go through the line on the defense. The plan of waiting until it is seen where the run will be made and then running behind his line to help, if the play appears to be on the other side, is disastrous to a good defensive game. It not only is dangerous, because it leaves the way clear for a splendid run on a double pass, but it is also especially harmful because it gets the tackle into the habit of waiting for every play to become well started, and this is fatal to a strong defense. If the play is around the other end, the tackle should follow the runner around and try to overtake him. It is sometimes possible for a fast runner to do this when he breaks through quickly. In following the man with the ball, the tackle must be on the watch constantly for a double pass. If he suspects one is to be made, he must be sure not to be drawn in or blocked as he runs behind the line. It would be better, in that case, to go straight through. The tackle can do more to defeat a double pass than any other player, for, if he plays his position well, he will meet the runner when there is not more than one interferer to combat. If he then does not tackle the runner, he can force him to run so far back of the line that the rest of the team will be able to come to his assistance before he circles the end.

When the opponents are going to kick, the tackle has an especial burden resting on him because he is in a very advantageous position for breaking through quickly and stopping the ball. No other rusher should reach the fullback so quickly, unless, perhaps, the guard, because none other is so well placed and at the same time interfered with so little.

He should, therefore, go through with all his strength and speed, and jump high in the air to stop the ball. His hands should be raised at the same time in order to place as high an obstacle in the way of the ball as is possible. The tackle on the same side as the kicking foot has a better chance to stop the ball than his companion on the other side, and he must, therefore, put forth his utmost efforts. Frequently, the tackle, like the guards and center, can work some clever team play in conjunction with an extra man, whereby one or the other can go through the line with little opposition.

There are a variety of tactics which can be employed

in getting through the line, and every tackle should be able to use them at will. Those are best which enable the tackle to get through quickly and at the same time permit him to watch the runner closely. This is a point which ought to be deeply impressed on the minds of all the rushers. The situation changes so quickly when a run is being made that it is not safe to have the eyes off the runner for a second. The methods usually employed in breaking through the line are : striking the opponent in the chest quickly and hard, and following it up with a shove to one side when he is off his balance; whirling suddenly around him, using either foot as a pivot; ducking quickly to one side ; making a feint to go one side and going the other ; striking the opponent with the head or shoulder and lifting him aside ; stepping a little to one side as the opponent comes forward and swinging him through behind him. The tackle can sometimes secure an advantage for breaking through by pushing his opponent back from the line just before the ball is snapped. He must be very free to move, and go through with a jump. It is better to keep as low down as possible in doing this.

The position which the tackle should take on the defense against mass plays from the center of the field is shown in the diagrams further on. He should move off from the guard sufficiently to protect the side of the field and at the same time be able to spring back close to him on any play directly forward. It is his special duty to tackle the runner if he comes out at the side of the formation. In case the runner does not come out before the opposing rushers meet, the tackle should dive in and secure him, if possible, but in doing this he must be careful not to leave too great a space between himself and the guard, as an opening through which to send the runner may be intended at that very point.

It is impossible to lay down rules of action for the tackle on wedge plays in the line. He must work according to his best judgment based on the situation ; but an important factor in successful play will be to put in the work low down. If he is caught by the wedge in an upright, or nearly upright position, he will be rendered absolutely useless. For this reason, it is often best to dive in at the side of the wedge about knee high and try to tackle the runner, or cause him to fall over him. If the wedge is revolving, it is often best for the tackle to fall down in front of it. The tackle must consider it his first duty to assist the center and guards in checking the wedge, and leave the other players to attend to the runner if he comes out from behind or at the side.

On the offense, the tackle cannot leave any unprotected space between himself and the guard, if it be occupied by an opponent. He must therefore always take the inside man. This may require him to play close to the guard. From this position he must do all his running with the ball, all his blocking, all his interference for the runners, and make all his openings ; varying his attitude toward his opponent to meet the special need of the moment. In making his opening the tackle has to outwit and combat a very free opponent, one who, as a rule, is constantly changing his position. This renders it difficult, sometimes, to make an opening because frequently it has to be done while the opponent is changing his position, and when, perhaps, the tackle himself is not in a favorable position for making that particular opening. Likewise, when trying to block his opponent, the tackle must follow him closely and keep in front of him, and must be all on tiptoe to dart forward to get in a body check before the opponent acts.

When the tackle runs with the ball or moves away from his position to accompany the runner, he is much more at liberty in choosing his place in the line. His great aim should be to take a position which should not be noticeable by its strong contrast to previous ones, and yet, at the same time, be one which he can use to the greatest advantage in the play in hand. Usually that position should be up in the line not more than two or three feet from the guard, but sometimes it is better to stand a little behind the line.

It is most important to the tackle when he runs with the ball that he get away from his opponent with the utmost quickness, and then, that he run with tremendous speed and power. The secret of successful running from any position lies in this. The practice given to improving in this particular should be faithful and constant. The run of the tackle cannot be successful until there is added to the quick start and strong headway, such training in taking his course that he will neither run too near the line, nor too far back from it ; and the ability to circle around the quarter-back and take the ball from him without a diminution in speed, and then plunge into his opening with a force which cannot be stopped short of several yards. Much depends on the course taken. The tackle's failure in running often results from slowing up to turn into the right opening and thus losing his power. Instructions in running and holding the ball are given in the chapter on the half-back and full-back.

THE END-RUSHER.

The end-rushers fill two of the most important positions on the eleven. In defense, their especial duty is to prevent the long runs of the game. It is an unusual thing for a long run to be made through the center part of the line on account of the support given the rushers by the quarter-back and half-backs. Let a runner once get around the end with one or two interferers ahead of him, as is usually the case when such runs are made, and he is likely to go a long distance down the field and not infrequently make a touchdown. In defending his territory against these runs the end stands at the most remote part of the field for assistance to be rendered him. He is at the extreme part of the rush line and has no one close to him to help him. His nearest neighbor, the tackle, must be depended on for most of the assistance, and when he cannot render it, the end is put to the test of tackling a runner preceded by a group of interferers. In such an emergency a deep responsibility rests upon the end-rusher, because he is probably the last man left to prevent a long run and perhaps a touchdown, producing a sensation akin to that of the full-back when he alone stands between the runner and the goal.

Moreover, the end-rusher has to meet the runner under most trying circumstances. The runner and the interferers have gotten well under way ; they have passed the most dangerous spot in the line and are coming on at great speed. The interference is now more focused and effective in arrangement than it has yet been. There are more interferers and they are more closely bunched. At the same time, the end well knows that he is an especial mark on all sides. He realizes that a particular man is appointed to do his utmost to check his play and that if this man fails to do it, the work is to be attended to by the other interferers who come immediately after. Under these difficulties in tackling and maneuvering, it is not strange that every captain is most careful in the selection and training of his end men.

The kind of man who could play a brilliant game at end, might not, perhaps, be able to fill any other position in the rush line, yet this is not necessarily true. His qualification would be questionable only as regards build and weight. There are most brilliant end players who only weigh about one hundred and fifty pounds, and sometimes a little less, but the tendency now is toward selecting slightly heavier players for that position in order to gain more weight with which to meet the tremendous on-rush of the interferers. But it is not infrequent that the light, agile, cat-like men are much more likely to tackle the runner, and so are selected in preference to those possessing plenty of weight but less skill. The tackling of these light, quick men is necessarily most brilliant, because they do not bore their way through to the runner but seize a momentary opening to put in their telling work. Such a man, as has been said, could not play in any other position in the rush line, for he would not be heavy enough to stand the hard pushing and plunging to which, for example, the tackle is subjected. With the exception of meeting the end plays and plays between the end and tackle, the end-rusher does not have the hard, wearing work of the other rushers. Not that he does not have plenty of work to do, but he is not constantly combating an opponent and struggling with might and main to get through the line, thus being subjected to the little knocks and bruises which the other rushers have to endure.

The end-rusher is at liberty to take any position he chooses on the offense. His one thought, however, should be to take that position from which he can best operate in helping out the play. Many end-rushers fail to do this. Some ends play up in the line and follow their opponents wherever they move, no matter how far out they go. Others take a stand a little back of the line, about a yard or two from the tackle, shifting this now and then as the play suggests and admits. This latter is generally the best position which can be taken for helping in the interference, and it is also a better position from which to start if the end-rusher is to run with the ball himself. Whenever the end-rusher is going to take the ball he should carelessly assume a position a little nearer the quarter-back — perhaps almost behind the tackle. Otherwise, the distance which he would be obliged to run before he reached his opening would be so great that the opponents would have enough time in which to intercept the play. On this play the quarter-back should give the ball to him by a short pass and then run ahead to interfere.

If the end-rusher plays up in the line he should always take the inside man when acting on the offensive. This is a point frequently forgotten, and oftentimes is the reason why end runs are stopped before the runner reaches the end. The end-rusher should also remember to help the tackle whenever the latter takes the ball. In this case it may be necessary for the end-rusher to step in and block the opposing tackle, but if the tackle can break away from his opponent without assistance it is better that the end should follow the tackle right around. When the tackle is to go into the line the end can do no better than place his hands on his hips and steer him into the opening. If the end-rusher does this well he can be of great assistance to the tackle in running, and at the same time prevent him from being caught from the rear. The best way to play the end position in making the different evolutions is shown in the chapter containing diagrams.

On kicks into touch the end-rusher must cover the ball well and secure it the instant the full-back puts him on side. Whenever an opponent secures it the end-rusher on that side must be on the watch to prevent his quickly putting it in play at the point it crossed the line. He should also be on the watch for all side-line tricks. The other end man should return quickly to his position to guard his field against a throw in from the side or any quick play. The end-rushers must be sure to keep their eyes on any outlying men who might receive the ball on a pass.

"Be the first man down the field on a kick" is the motto early instilled in the would-be end-rusher, and to do that and be there in time to tackle the catcher before he starts is no small accomplishment. It means that with a good punter, who has perhaps the wind behind him to propel the ball, the end must be exceedingly quick in starting and very swift of foot. If the end fails to get down the field in time, the ball will be carried or kicked back, whereas a swift runner might be able to prevent this. Moreover, the full-back ought not to be compelled to limit his kick because of the slowness of the end-rusher.

It requires long practice and much careful study to determine just the direction the ball has taken almost at the moment it is kicked without wasting time in turning around or in looking over the head into the air. Likewise it requires practice to decide upon the best way of approaching the man to whom the ball is kicked. It is a common fault for end-rushers to run blindly down the field without knowing the exact direction which the ball has taken, when a little study of the faces and actions of the half-backs will indicate in a second whither the ball is going.

Another common fault with the end-rusher is the failure to tackle the man who gets the ball. This results largely from over running him. The player with the ball simply jumps to one side at the proper moment and lets the end go by in his headlong run, and then goes down the field. The one remedy is that he should slacken speed a little as he approaches and watch for a chance to tackle.

Care should be taken by the end-rusher as he runs down the field to approach the player who has received the ball so that he will be forced to run on the *inside* of him. Then, in case the end misses his tackle, he will fall into the hands of the other rushers, now near at hand. The position of the end-rusher when a kick is about to be made, should be such that he can protect the field. Usually he draws off well from the tackle. This must be done without fail when he has a large field to guard, that is, when the other end of the line is near the side of the field. The general form of the rush line as it advances when a kick is to be made, is described in the chapter on team play.

It may be said further, that usually the end-rusher should start his line of direction slightly towards the side lines until he gets the first inkling of the direction the ball has taken. He should then bear in or out still farther, according as seems best. This would not be good advice to the end-rusher who stands close to the side line. The reason for the end taking such a start is that he should protect the whole field against a run, and the least protected part should be attended to first. This suggestion has especial weight when there is a great deal of space between the end-rusher and the side line.

The end-rusher must be especially watchful at the start for signs of a short kick, or for one which goes to the side. Sometimes these are caused by inaccurate kicking, or by the partial stopping of the ball by an opposing rusher. In any event, he must be careful not to over-run the ball, and must secure it whenever an opponent puts him on side by touching the ball. If the end is in doubt where the ball is, he should glance around quickly and find out. The end-rushers must be especially careful when the ball is kicked from near the side of the field, for it often happens that only one end can be near the opponent when he catches.

The end-rusher should be under the ball when it falls, and if the opponent is a good catcher he should usually force him to make a fair catch. If, however, the end-rusher is where he is absolutely sure of securing the catcher if he should run, it may sometimes be better for him to give the opponent a slight chance to run for the sake of increasing his liability to drop the ball. This liability is further increased by a hard tackle just at the moment the catcher starts. The end should be on the watch to secure the ball at such times. He should also make sure that the catcher does not pass the ball to a companion near at hand.

There are many conditions to be met by the end as he goes down the field on a kick which cannot be described. He must note them as they come and act accordingly. One of the hardest of these is to know how to handle bounding and rolling balls. Observing the angle at which the ball descends, also the way it acts for two or three bounds after it strikes, will give some information on which to base action, but there is a constant uncertainty ; and in those cases where the ball is revolving on an axis constantly shifting as it goes through the air, there is no certainty of its action after it strikes the ground. It therefore takes the most careful playing at such times on the part of the end-rusher, for one of the opponents may dart in opportunely and seize the ball and go sprinting up the field. If there is any chance for this, and he is not well supported with helpers, the end-rusher should immediately touch the ball and force a down for the other side. Furthermore, when a kicked ball is likely to go over the line in goal, the end-rusher should do his utmost to touch it just before it reaches the five-yard line so that it shall be down at that spot and shall not be brought out to the twenty-five yard line.

THE QUARTER-BACK.

As popular opinion has always assigned the snap-back's position to the largest man on the eleven, so likewise has it given the quarter-back's position to the smallest man. There is less reason in having the smallest man quarter-back than the largest player at center. Indeed, there is no question that a swift, agile man of one hundred and sixty or one hundred and seventy pounds would be the most useful quarter-back, if his other qualifications are equal. The trouble is that the man of such a weight, who was qualified to fill the quarter-back's position, would be the man who would be most needed at tackle or end, or as a running-back. There is rarely more than one man with these qualifications on the best teams, while there are usually several men of sufficient speed and agility among the candidates, who perhaps could not be useful in any other position, and yet are too skillful players to loose. The result is that on university elevens the quarter-back is usually a man who weighs from one hundred and forty to one hundred and fifty-five pounds, is agile and swift, is a hard worker, with great endurance and unlimited pluck. Well does he need all of these qualities, for he must always be in the thick of the fight. No play can take place from a scrimmage without his being a medium in its execution, not only in the passing of the ball, but also, if he does his duty, in assisting the runner on his way up the field. Not that he runs ahead of the runner every time, for he is unable to go in front on some plays, but he can always get behind to push if the runner is stopped, or to block off those who try to tackle him from the rear.

The quarter-back's position demands a peculiarly heady player at the same time that it calls for agility and quickness. No other player on the eleven is forced to do as much thinking and planning while in the midst of most skillful and invaluable work. He has no chance to " soldier," either mentally or physically, as the rest of the eleven may do, to a limited extent, occasionally during the progress of the game if so disposed. His brain must be as clear as his muscles are quick and steady. He has to translate with absolute exactness every signal which is given, and as accurately carry it out by forwarding the ball in the most advantageous manner possible to the player who is to receive it. On no account, then, must a man be selected for this position who is inclined to become " rattled," for the position itself is enough to render unsteady the coolest man.

When the quarter-back is appointed to give the signals for the play a new duty emphasizes the importance of his being a heady player, for he then is made the general of the game. By having this duty to perform the chances for his making a mistake in giving the ball to the wrong player are perhaps slightly decreased, but the demand for clever judgment and shrewdness in field tactics more than offsets this.

The quarter-back must know no physical fear. He must be fearlessly unconscious that there are several opponents almost within reach of him who are doing their utmost to fall upon him. No nervousness must enter into his work ; else he is not the man for the position.

In assuming his position on a down, the quarter-back is allowed considerable freedom. Some players prefer to receive the ball close up to the center-rusher and then move away as they pass it on to the runner ; others take a position between the two, just as far away as is possible while still being able to reach the center conveniently for giving the signal.

The quarter-back who plays close up to the center renders himself liable to be interfered with in his pass by the opposite center and guards, who may reach over to check his play ; at the same time he cannot so well take part in the interference on end plays. On the other hand, the quarter-back who takes his position far behind the center is limited in some of his plays. He can be of more assistance, perhaps, in helping on the end plays, but it will be impossible for any of the guards and tackles to run with the ball with any chance of gaining ground, because they will have to run so far behind the line to receive the ball that they will easily be tackled. When the quarter-back takes this position he will have to give the signal in some other way than that usually followed. It has been customary for the quarter-back to press the calf of the center rusher's leg, or some other part of his body, with his thumb when he is ready for the ball; but there are reasons why some other signal would be better at times, and the giving of the signal would be of little moment if there is to be a decided advantage gained by playing so far behind the center. It is accepted as the best way for the quarter-back, in playing his position, to stand bent over, at arms length from the center, with his eyes fixed on the ball.

He has already learned the position of the player who is about to receive the ball as he glanced around at his team when the signal for the play was given. The instant that he gives the signal for the ball to come back he turns quarter round, throwing his right or left foot well behind for a brace, according as he wishes to pass the ball to the right or left. The quarter-back must not take his final position for receiving the ball before the signal for the ball to come back is given; otherwise the opponents will have time to study out his method of passing for the different plays and can guess in what direction the run will be made. It is all done so quickly in the other case that there will be no time to anticipate the play.

The quarter-back should never give his private signal for the ball until the captain has given the signal for the play, and then only after he comprehends it himself. In a well drilled eleven the quarter-back understands the signal for a play the instant it is given, and yet it is not a rare occurrence in important games for signals to be mixed or the key numbers to be left out. In that case the quarter-back should not signal for the ball until the signal for the play is made plain or a new one given. It is now a common practice for the quarter-back to give the signals for the play himself, whether he is captain or not. This has grown out of the fact that he is in one of the best positions for observing the whole field, and also because he will no longer need to interpret the signal after it is given, but can call for the ball as soon as he thinks best. This facilitates the play somewhat and lessens the liability of making mistakes in translating the captain's signal.

There are three styles of passing a ball used by quarter-backs. Two of these make use of only one arm in forwarding the ball—one by an overhand and straight-arm movement especially valuable for passing long distances, but too slow for ordinary use; the other by an underhand pitch with an easy, natural swing of the arm. This latter style is the quickest of the three, for no time is lost in raising the arm into a position for delivering the ball. This pass supplements the movement of the ball along the ground most quickly and naturally. In the third style of passing both hands and arms are used and it is closely allied to the one-arm underhand pass. This insures accuracy, but places limitations on the distance the ball can be thrown. It is commonly used in all short passing. It would be of great advantage if a quarter-back could pass accurately with either hand.

In receiving the ball from the center the quarter-back should stop it with the hand which corresponds to the leg already placed behind for a brace and immediately adjust the other hand to it for a pass. This is done by placing one end squarely in the hand from which the pass is to be made and spreading out the fingers. The hand should then be bent at the wrist until the ball rests against the forearm. The ball is now in a position for a pass. Care should be taken to have the hand squarely behind the ball, also to have the long axis of the ball parallel with the forearm. The easiest way to make a long pass is to swing the arm at full length just below the level of the shoulder.

The quarter-back must need give considerable time to practicing all parts of his work in receiving, handling, and passing the ball. It is no easy matter to receive the ball as it comes bounding back from the center-rusher and adapt it to the hands for accurate passing while quickly turning into position to deliver it to the runner; but it is necessary for the quarter-back to do this in order not to be interfered with by the rushers who break through the line, and also not to delay the runner. It requires long practice, also, to be able to handle the ball and be off the instant the ball is in the hands, but it is an achievement which enables the quarter-back to be of great service in end interference. Unless, however, there is the most skillful handling of the ball it is impossible for the quarter-back to get ahead of the runner without delaying him. It requires much practice to be able to do quick and accurate passing — to be able to place the ball at just the right distance ahead of the runner and at just the right height and at just the right speed, so that he shall not be delayed an instant, and can give his whole thought to running and dodging.

Too great stress cannot be laid upon quick work by the quarter-back. It means success or defeat to some of the plays. At the same time the quarter-back must be exceedingly careful in handling and passing the ball. It is better to be a little slow than to be quick and unsteady. He must never become excited and lose his self-control, for that would be disastrous to all careful work and also would be likely to cause him to make mistakes in signals.

On all dashes through the center it is better for the quarter-back to make short passes of the ball at the runner's waist. The ball must not be passed fast and it must be most *accurately* placed, for the runner is bent over for a plunge and is not in a position to handle it, unless on a slow and accurate pass. These points are worthy of the most careful consideration, for much of the fumbling by the half-backs is due to poor passing. What would ordinarily be an excellent pass if the half-back were at some distance, would be a poor one when he is coming forward at full speed, with his body somewhat bent at the waist, and his attention partly on the ball and partly on the opening he is to take. In this case, also, a high pass is harder to catch than a low one, because the hands will have to be raised quickly from their position at the waist.

The quarter-back should also use the greatest care in his pass to the full-back for a kick, for a poor pass will most likely result in the opponents stopping the kick and securing the ball on four downs, if not on a fumble. The full-back can kick most quickly when the ball is passed at his waist.

Some quarter-backs prefer to *hand* the ball to the runner as he dashes by, whenever that is possible. This method, without doubt, is best when the guard or tackle runs around for a plunge through the line between center and guard, or guard and tackle, on the other side of the center. In this case the quarter-back will turn half around, with his back to the center-rusher, the ball being held by the ends between the extended hands. In most other cases an advantage is gained by *passing* the ball, because the quarter-back will not be in danger of being tackled by the opposing rushers or quarter-back, as they break through the line, and also because he will be free after his pass to give his whole attention to helping the runner. He may do this either by going through the opening and pulling the runner after him ; by grasping him and going through with him; by shoving him hard when he strikes the line; or by jumping into an opponent who has broken through in the path of the runner. Occasionally it may be better to hand the ball to the runner when the quarter-back runs out to the side to interfere for him; but even in that case, a short pass usually facilitates the play because the quarter-back can run faster and do better interference when free from the ball. It is of great assistance in getting into the interference on end plays for the quarter-back to be able to pass the ball accurately on the run, for every fraction of a second counts in making a helpful connection.

On the defense the quarter-back usually hovers in the rear of the center and guards, watching his opportunity to go through and tackle the opposing quarter or half-backs.

A powerful style of defensive play has now, however, been largely adopted, in which the quarter-back takes a position behind one of the tackles, while a half-back is brought up to a corresponding position behind the other tackle. They there await the play without attempting to go through on the instant the ball is snapped, and as the line of their opponents separates for the play, the one on whose side of the center the opening is made dives into it to meet the runner before he can strike the line.

He must know just when to go through the line and when to wait in order to see where to meet the play; also through which opening in the line to go in order to best check the play. Some shrewd guessing can be done, which will help determine this by noting all the signs of the direction of the play spoken of in the chapter on team play. The center and guards, and sometimes the tackles, should help the quarter-back find his opening and assist him in getting through. The quarter-back should always be helped through when the opposing team is going to kick, since it will be much easier for *him* to go through quickly on account of his size and quickness in starting. If the rushers and the quarter-back work together on the defense the latter can be a most valuable adjunct to their play, because he is free to move anywhere. When a runner is checked or tackled, the quarter-back, as indeed all the eleven, should endeavor to pull the ball out of his hands before he calls " down." The quarter-back often has a good chance to do this when the runner is entangled in a mass.

THE HALF-BACKS AND FULL-BACK.

The half-backs and the full-back, who is practically a third half-back, stand usually from two to four yards behind the center of the line. They group themselves at short distances from one another and in a way to best assist in carrying out the play which is about to be made. There is a difference in the latitude given the half-backs and full-back on different teams in arranging themselves for each play. Some captains require these men to occupy the same position on every play, claiming that it is of great advantage in obscuring the play to have a fixed arrangement. On other teams the half-backs and full-back are allowed to move about, and shift their places to the position in which they think they can best help out the play.

There is also a great difference among teams in the placing of the half-backs and full-back in reference to *each other* and also in reference to the rush line. In general, the full-back is stationed behind the center and usually about a yard or a yard and a half further from the line than the half-backs. On some teams, these three play close together, separated by not more than a yard or a yard and a half; on others, they are separated from two yards to three yards and a half. There is also a decided difference in the distance behind the line which the backs play. This varies from two to five yards.

The arrangement of the backs should, in a measure, depend on the style of game to be played; and the style of game should be determined by the composition of the team. That is to say, that if it is deemed wise to play a center game, it can best be done by bunching the backs ; while, on the other hand, the combinations can be best made for an end game when the backs are more spread . apart.

Captains who are limited in the selection of their play-ers will find it well worth their while to consider the arrangement of the backs, both in regard to their relative distance from each other, and also in regard to the distance which they stand behind the line. Indeed, there is an opportunity for fine generalship in deciding upon the place for these ground gainers.

When the three men who are to occupy positions behind the line have been decided upon, there is also need of careful consideration in determining which posi-tion each one of the three shall fill. The full-back is usually selected for his ability to kick, and yet, it is some-times better that the man occupying that position should act as a half-back until the signal for a kick is given, and then drop back ; while a half-back sometimes could do more effective work in the middle position during the general play. If one of the backs is slow, his best posi-tion is usually at full-back, for there he receives the greatest protection and help. The light, quick men can succeed better at half-back than the slow, heavy men.

It frequently happens that one of the backs invariably carries the ball under the right arm and is able to use only the left effectively in blocking off, or *vice versa.* This fact should be considered in determining which position the men shall occupy.

It is unfortunate for a half-back to be so limited, but many of them are, and they do not practice with the other arm enough to train it. Some naturally run in one direction better than in another ; or some are surer and stronger of foot, perhaps, when running around on a particular side. A player is sometimes put in the right or left position because the interference is stronger on that side ; or possibly the arrangement is made to take advantage of a certain known strength or weakness in the team which they are to meet.

The half-backs and full-backs are largely the ground gainers for the team and most of the advances into the enemy's territory are made by them. For this reason, only men who possess special qualifications are selected to fill these positions. In quickness and agility they should equal the quarter-back ; in point of speed, ability to dodge, courage, and dash, they should be unequaled by any man on the team. Again and again they must rush headlong into the line, oftentimes only to be hurled back by the opposing rushers who plunge through upon them. Yet, never losing courage, again and again they must come to the rally, now attacking the opponent's center by heavy plunging now trying to make a detour around the wings.

Too great emphasis can not be placed on quick starting. The inability to get under headway quickly is very often the difference between a first-rate half-back and a second-rate one. The second-rate half-back may be just as fast a runner, and may be just as hard to stop when once under way, but he does not get under headway nearly so often, because he loses so much time on his start that he is tackled before he passes the critical point in the run. On all plunges into the line the utmost speed must be used in conjunction with the quick start. The distance is very short in which to get under headway, and there is need of the greatest force to project the runner through the resistance, as well as need to reach that point of resistance in the shortest time. It is common with many elevens to have one heavy back to do the plunging into the line, but frequently this man is so slow in his start that he is not so effective for line-breaking, against a strong defense, as the lighter man would be. It very frequently happens that in choosing the half-backs, men have to be selected who have only part of the qualifications for the position ; who perhaps can run fast, or, again, are what are termed "fighters," but lack some of the other requisites When such is the case, the captain should immediately take means to train these men in the other necessary qualifications for good half-back play. It is indispensable that a half-back should be able to run into a line *hard* time and again, and with no fear or hesitation. It is likewise most necessary that a half-back should be a powerful runner and not easily stopped ; one who does not fall easily but keeps his feet well when tackled, and struggles on for the gain of a few feet. But he would be a much more useful man if, at the same time with this pluck, determination, and ability to stand on his feet under difficulties, and keep struggling forward, he also had the ability to dodge an opponent or ward him off with the extended arm, instead of running straight into him.

Dodging in running can be cultivated through the study and practice of its points of deception. The underlying principle is the quick movement of the body, or portion of the body, from a point where it would have been if it had continued in the same direction. In the most simple form of dodging the runner suddenly changes his direction. As usually practiced, the runner is obliged to slow up a great deal, in order to change his course. In all dodging, the runner, if at topmost speed, must slacken speed a little, just before he reaches the tackler, in order to reduce the size of his stride so that he may have a proper balance for projecting the body in another direction, or so that he may make certain preliminary body motions which cannot be made when at full speed.

There are several ways of dodging, but one man seldom possesses more than one or two. The zigzag dodge, which used to be so common when individual running and poor tackling were in vogue, is performed by a combination of leg and body feints. Its weakness is that it retards the runner too much. In another dodge the runner strides suddenly one side with a long step. This is a very effective method for long-legged runners. In another, the runner sways his body from one side to the other, the legs being planted wide apart as each step is taken in a zigzag course. The runner moves in the same general direction until the opponent is reached and then darts to one side. Still another dodge is made by drawing the hips away, and in this dodge a clever use of the arm is valuable. It is one of the most effective, since the hips are usually the part aimed at in tackling. Another way is to duck under a tackler by bending the body low at the waist. This is practiced most effectively by small men and is most valuable against high tackling. Another method is to turn the body completely around when about to be tackled, upon one foot as a pivot. This comes into splendid use when the tackler has been unable to grasp the runner with both hands. In another form of avoiding a tackler, the runner, on being approached from the side, slows up a little; whereupon the opponent delays just long enough to allow him to go around by putting on a burst of speed.

Good dodging is not complete unless there is added to it the power to use the arms well in warding off. The latter supplements the former most effectively when well done. When the tackling is high, or when the runner is well bent over, the arm should be extended against the face or chest of the opponent. Often, on a long dive or reach for the hips by the tackler, the runner can break the hold by striking down with his arm. All the above styles of dodging can be acquired by practice. It is better to practice them with only one or two men to act as opponents, after the movement has been learned.

There is another requisite needed by the half-back in addition to dodging, and that is the ability to follow an interferer or interferers well. Half-backs differ greatly in skill on this point. The work of escaping a tackler should not rest wholly in the interferers' hands, as it so often does. The half-back should supplement the latter's work by taking advantage of the protection given him to work every ruse and feint he knows. Where there are several interferers, there is a chance for the runner to move from one to the other as occasion suggests. It needs quick wit and agility to follow interferers well, but much can be learned by practice with or without opponents, and every half-back should devote himself to perfecting his play in this particular.

The half-backs must be good catchers, not only of kicked balls, but also, and especially, of balls passed from the quarter-back. Oftentimes, the fault of a muff or a fumble can be laid to a poor pass, but if the quarter-back is unsteady on his part, there is all the more reason that the half-backs and full-back be skillful catchers. If

5

weak in catching, much practice should be given by the
half-backs to perfecting themselves. They should work
at this in conjunction with the quarter-back in order that
they may get used to each other. In catching short
passes, it is usually better to catch the ball with the
hands. This is surer because the hands can adapt themselves much better than the arms to the position and
shape of the ball when a man is running. In running
sidewise to the pass, as it is necessary to do in so many
plays, the arms could not be used without checking the
speed ; while there need be no diminution in speed when
the ball is caught in the hands, provided the quarter-back
does his work well.

There are three ways of carrying the ball, and each
has its proper occasions for use. When the play is
straight through the center the general order to the half-
back is to put the head down on a level with the waist,
gathering the ball up under the body with both arms, because there could be no use for an arm to ward off an
opponent until the line has been penetrated, and there is
great danger of losing the ball by the pulling and haul-
ing to which the runner is subjected. After the runner
is well through the line and has a chance to run freely, he
should transfer the ball to the side of the body opposite
the arm with which it is necessary to ward off. The
runner should look for opponents as he emerges from the
opening, and likewise for interferers. Where the play is
through the more open part of the line the runner should
usually carry the ball under the arm which is away from
the opponents who are likely to meet him first, shifting it
to the other arm when necessary. - In this case, likewise,
it is occasionally better to carry the ball in both hands
until there is need for warding off an opponent, at which

moment the ball can be easily shifted to whichever arm it is desired. This provides for any emergency. This way of carrying the ball is especially valuable in dodging, since the ball can be placed quickly under either arm and a better defense made ; for if forced to dodge, the runner may transfer the ball to the arm away from his opponent and have the other free to ward off. By moving the ball from one side to the other in front of the body while running, the dodge will be made more effective.

In carrying the ball under the arm it should be held well forward, because it can be held more tightly in this position. The reason why the ball is often pulled out from under the arm is that it is held so far back that the strong muscles of the chest are of little assistance. When held in this position the ball is often forced out from under the arm when the runner is thrown to the ground. By testing these two positions it will be easily seen which is the safer way. If a runner is inclined to lose the ball he should practice squeezing it in the most approved manner until he has trained himself to hold it fast under all circumstances.

We have already spoken of the runner getting under headway quickly. It is also necessary that he should run with all his speed ; whether he plunges into the center part of the line or follows the interference out to the wings (unless he is obliged to slow down in order to receive the ball, to let a runner in ahead of him, or to get by an opponent). No runner is so invincible in all his play as he who rushes with all his strength ; who shows by his every movement the determination and power with which he is charged ; who inspires in his opponents a hesitancy and dread of tackling him ; who never gives up when tackled but keeps struggling on, twisting, squirming, and wriggling himself out of the grasp of one after another until he can no longer advance. Such a man is worth a dozen who hesitate.

The dashing runner is the one who usually makes the advances. If he goes through an opening he goes through on a jump. Such a man, when checked, will keep his feet and legs going like a treadmill and will bore his way through in spite of resistance. This sort of pushing accomplishes wonders. For effective application of power it is worth vastly more than the same amount of force applied slowly, for the attack is sudden and continuous. Its effectiveness, however, is altogether dependent on the head being well bent over, so that the whole weight and impetus of the body is forward, for the legs are then in a position to exert the greatest power.

Another reason for running into the line well bent over, is that it is much more difficult to tackle a runner when in that attitude. It is impossible to get under a short man in order to make a low tackle when he is coming straight toward one, and the result is that the tackler receives the runner's head in his stomach, or if he be good in the use of his arm, he will very likely have a hand thrust into his face or against his chest. At such times, the runner is very often able to slip past.

Again, running with the head down enables the runner always to fall forward when tackled. This usually means a further gain of two or three yards.

In running low care should always be taken not to lose the balance. After considerable practice the balance can be very well kept when running much bent over and still great speed be maintained. As soon as the line is cleared and there are no opponents very near, the runner should assume a more upright position so that he can run at his utmost speed, lowering his head whenever he thinks best.

In making the end plays, the runner need not put his head down except, perhaps, when it is necessary to duck under a tackler. He must now put on speed up to the full limit of the interferers, following them very closely, now using this one and now that, according as the danger shifts. He must constantly be on the alert for changing his position to take advantage of every little help, or to prevent being pocketed, at the same time being ready to break away from his interferers if he sees he can gain more by so doing. In general, the runner should keep behind his helpers until the last, but now and then an opportunity comes which he ought to accept.

The light-footed, agile man who can keep his balance well is physically best capacitated for running behind interferers. To do it well the runner should be able to change his stride to meet the emergencies which arise in passing from one interferer to another, or in following very close when a long stride would cause him to stumble over his interferers.

Another requirement which the backs, or at least one of them, presumably the full-back, should have, is the ability to kick. It would be well if all three possessed this ability, for there are times, now and then, when consternation could be brought to the opponents by the half-back returning a kick. But this could happen only occasionally, and it is much more important that the half-backs be especially strong in running with the ball, for that will be their main work. The full-back however, should be a skillful kicker both in punting and drop-kicking.

It requires long practice to punt well. The oval shape of the ball precludes simply tossing or dropping it from the hands and then kicking it, to get the best results.

The mechanical construction and adjustment of the

muscles of the leg and body in their relation to kicking require careful study. Long practice is necessary to be able to regulate the power, and at the same time determine the angle and direction which the ball shall take. All the practice which the full-back can get to acquire skill in punting will be well repaid, for it will make him of inestimable value to his eleven.

Where the full-back does not know how to punt, the following directions will be found helpful : Hold the ball between the hands, the ends pointing to and from the body, lacings up. Extend the arms horizontally in front and bend forward with the body until the ball is held just below the level of the waist. Take a short step forward with the foot not used in kicking, and at the same time drop the ball from the hands and bring the kicking leg quickly forward to meet the falling ball about knee high. Do not try to kick hard at first. Attend simply to dropping (not tossing) the ball without changing the relative position of the axis. This must be closely regarded or there will never be any certainty as to where the ball will go. The first point noticed by a novice will be that the ball reaches the ground before his foot meets it. This shows that the foot was not started forward soon enough. One way to obviate that difficulty is to drop the ball from a higher point ; but the best point has already been selected and the tardy member must be trained to be on time. It will also be noticed that sometimes the ball will meet the leg above the ankle. The aim should be to have the ball fit into the concave of the extended foot, and it will probably be necessary to give the ball a slight toss forward in order to make the kick powerfully. Care should be taken when doing this that the ball is not turned, or tossed so far that power is lost. In practicing in this way it will at first be noticed that the whole force of the blow will be given by using the leg from the knee down. This, one can readily see, would weaken the blow because the leverage is short and the muscles which extend the lower leg not especially powerful, and at the same time it is very trying to the knee joint. The most powerfu kick would be one which had the leverage of the full length of the leg, thus bringing into play the strong abdominal muscles to add speed and power. In making this kick, the leg should be extended at full length (with toes pointed) and should swing on the hips as an axis. After the forward kick has been learned so that it can be well executed, the side kick may be attempted. In this case the ball is dropped a little to the outside. The great advantage in the side kick is, that if not too much on one side, a very considerable increase in power can be gained, because a longer swing can be given to the leg, and because the swing is further assisted by some additional muscles which give increased power. Another advantage is that the full-back can take a step to the side and kick around an opponent.

In practicing, do not keep the leg rigid through all the swing. The muscles must be sufficiently lax to make the swing easy, the rigid contraction coming just before the foot reaches the ball.

The angle at which the ball is kicked can be regulated by elevating or lowering the point of the ball farthest away from the body, or by dropping the ball in such a way that the position of the foot in the arc described by it shall regulate the direction which the ball shall take. If the kicker wishes to make a high kick, he drops the ball so that the foot reaches it when knee high or above, and when he wishes to make a low kick he allows the ball to get closer to the ground before his foot meets it. By trial, it will be found that a point varying from about six inches above to six inches below the height of the knee is the place of greatest convenience and power.

After punting and drop kicking has once been learned, the whole practice should be centered on kicking quickly. The ball should be caught, adjusted, dropped, and kicked just as quickly as possible. In practicing this, it will be found expedient to have several balls for the quarter-back to pass. After practicing for a few weeks in this way the full-back will find that he can stand considerably nearer the rush line and still avoid having the ball blocked.

The drop kick is made by dropping the ball on one of the small ends and kicking it with the toe at the instant it rises from the ground. Some kickers prefer to have the ball lean toward them at a slight angle as it strikes, others to have the ball lean slightly toward the goal, and still others drop it with the long axis vertical. The latter style is most commonly used. Practice in all these will determine in which position the foot meets the ball most naturally. The ball should be kicked with a free and easy, though quick, swing of the leg. If close under the goal the kick may be made more quickly with a short half swing, whereas in punting the leg is swung from the hip and the large abdominal muscles of the body brought strongly into play. In drop kicking very accurate, rapid, and effective work can be accomplished when the swing is made almost altogether from the knee joint with only a slight swing from the hip. Beginners frequently make a great mistake in drawing the foot far back in preparation for a long drop kick. By extending the leg below the knee quickly and suddenly, so that the point of the toe will meet the ball at the instant it rises from the ground, great distance can be attained with little apparent outlay of force.

It requires a great deal of practice to be quick and accurate at the same time. The full back should place himself a little farther from his rush line in attempting the drop kick than in punting, because the ball starts lower and it is not so easy to control the angle it takes.

In trying for a goal from a place kick the ball should be brought out to a spot from which the angle to the goal and the distance from it are most favorable for the trial. If the touchdown is made directly behind the goal, or near it, the ball should not be carried far out into the field. A point should be selected where there will be no danger of the opposing rushers stopping the ball and from which it will be easy to kick the goal. Some men prefer to make the trial from a point not more than ten yards away, while others carry the ball out fifteen or twenty yards. The former always make a quick half swing of the leg in kicking, lifting upward with the foot as they kick; the latter usually kick with the leg swinging full and free from the hip.

The ball should be held between the outstretched hands of the quarter-back or some other player as he lies extended flat upon his stomach. The best way of holding the ball is to place the fingers of one hand behind it about three inches from the lower end, the fingers of the other hand being placed at a corresponding point at the top and slightly in front of the ball. The ball should be held in firm but easy balance, and the fingers should be so placed that it will be easy to turn it and least interfere with it when placing it down for a kick. Great care must be given to holding the ball steady.

When the spot has been selected from which the trial

is to oe made, and the player who is to hol1 the ball has prostrated himself in firm balance on the ground, at right angles to the line of direction, and on the right or left side of the kicker, according to the foot which he is to use, the ball being properly held between the fingers with the elbows resting on the ground, the kicker must proceed to sight the ball. He first asks the holder to turn the lacing of the ball toward him; next he tells him how he wishes the ball to point and at what angle, if any, using such expressions as " head forward" and " head up," meaning that the ball is to be tipped away from the kicker in the first instance and held vertically in the second. Other expressions like " head out" and " head in " indicate that the point of the ball is to be moved in or out in reference to the player holding it.

The sighting of the ball toward the goal can be done best by using the lacings as a guide, the holder being directed to twist the ball out or in, in reference to himself, by the expressions " lacings out," " lacings in." When the ball has been well aimed and everything is ready the kicker should tell the holder to " touch it down," at the same time moving forward to kick. In touching the ball down the holder must be very careful not to change the position. As the ball touches the ground the lower hand is removed in order not to interfere with its course. It is well to remove beforehand all pebbles or tufts of grass at the spot selected for placing the ball down, for a slight unevenness is often sufficient to prevent a goal.

The kicker should keep his eye on some point on the ball as he steps forward and aim to kick it in that spot. Practice beforehand will determine the best place to give the impetus. When the ball is vertical this spot will be found by trial to be very near the ground ; when the ball leans toward the kicker the best point for the kick is just below the lacing. The height of the point above the ground is nearly the same in both cases, but the point on the ball changes as the ball leans. If there is a wind blowing the kicker must take into consideration its force and direction in pointing the ball.

In catching kicked balls and long passes, it is usually better to catch them with the arms. Every effort should be made to take the ball when about waist high, for at that point the arms can be better adjusted to it. The body also, here much softer, can at this part be drawn in to form a sort of pocket, as it were, for the ball. Care must be taken not to have the ball strike high up on the chest, for it is then difficult to shape the arms well to receive it and the ball rebounds much quicker from its firm walls.

There are two ways of catching with the arms. In one, the arms work in conjunction with the body, the latter being used to stop the ball while the arms close around it. In this style, one hand and forearm should be held lower than the point of contact with the body, while the other hand and forearm should be held above that point. The arms should be bent and should not usually be extended far from the body. In the other case, the ball is caught entirely with the arms and hands. This can be done only when it is kicked well into the air. The arms are held parallel in front of the body about six inches apart, being half bent at the elbows and wrists. The instant the ball strikes, the hands are curled forward over it. The fault of catching in this way usually lies in the catcher failing to bring his elbows near enough together and so leaving a space for the ball to go through.

In nearly all plays the backs, from the nature of their duties, are among the first men to start. Their position behind the line renders their every motion conspicuous, and the watchful rushers upon the opposing team will be upon the constant lookout for some movement, glance, or position of the body that betrays the direction of the play which is about to be executed. On this account the backs should take the greatest precaution to conceal their intentions. It is of assistance sometimes in deceiving the opponents to assume a position as if being about to go in one direction when an entirely different move is intended, but if this is practiced too frequently it will defeat its own end.

EXPLANATION OF THE DIAGRAMS.

Before passing on to consider the following plays, a few words of explanation will be necessary.

The side of attack in every instance, when in their regular positions, will be represented by the solid dots (● ● ●), and the side acting on the defensive by rings (O O O). When it is desired to represent a player in a position other than that which he originally occupies the figures ۞ ۞ ۞ will be used. The broken line (— · — · — · —) will represent the course of the ball in the pass and the direction taken by the runner who receives it.

A simple dotted line (- - - - - - - - - -) will be used to indicate that a player is to *follow* the runner with the ball, while the solid line (———————) indicates that the man shall pass in *front* to act as a line-breaker or interferer. The arrows indicate the direction which the players shall take.

The men represented by the letters given in the diagrams are as follows : c, indicates the center ; q B, the quarter-back ; R H, L H, R E, and L E, the right and left half-backs and right and left ends respectively ; the right and left tackles are indicated by R T and L T ; while F-B represents the full-back.

It must be *distinctly understood* that the drawings are in a measure *diagramatical* and do not in all instances represent accurately the *relative distance* between the players.

For example : in the diagramatical representation, wide spaces are left between the individual men in the rush line, while as a matter of fact, when the game is in progress, the rushers stand so closely together that they can easily touch one another and are frequently placed shoulder to shoulder. This manner of representation has been decided upon as conducive to greater clearness in showing the relative positions and directions where a number of men are obliged to pass through one opening, and in case the beginner is misled by this in any way, his error will be readily corrected by careful study in other parts of the book.

In arranging the positions of the side acting upon the *defensive*, the quarter-back has been placed immediately behind one of the tackles while a half-back has been brought forward and stationed behind the other tackle. The abilities of the two half-backs should determine which position they shall occupy ; the points to be considered being the ability to catch the ball when it is kicked, and the qualification for meeting the heavy tackling in the line.

Sometimes it is preferable upon the third down, or when the ball is to be kicked, that the half-back stationed behind the tackle should *immediately* return to his proper position. At all other times the quarter-back and half-back usually remain directly behind their respective tackles as indicated, after the ball is snapped, until it becomes clearly apparent through which one of the openings the opposing side is to make their attack, and then to spring forward directly into this breach and meet the on-coming runner *in the line*.

This is considered a safer and more powerful defense than to have either one of these men attempt to break through, in the hope of meeting the runner behind his own line before he reaches the opening, and is the method adopted by the leading college football teams in the country. When opposed to a team using the running game almost altogether, *both* half-backs may be sent forward to support the line, the full-back alone remaining well behind the line for safety.

It will be noticed that the ends upon the side acting on the *offense* are placed near the tackles and are drawn slightly back from the line. We believe that the ends are in the strongest possible position for an attack in any direction when they stand about a yard and a half from the tackles, and about a yard back from the line. From this position they are of equal value in blocking, should the play be made around their end, while in plays through the center and around the opposite end, their position back from the line enables them to get into the play with far greater rapidity, and wellnigh doubles their efficiency. From a position *in* the line the running of the end, with the ball, which may be made a powerful play, would be extremely difficult.

Nearly every diagram represents *two* plays or more, and it should be borne in mind that, whereas in the diagram a play may be represented as made to the *left*, the same play may also be made to the *right*, and *vice versa*.

In representing the arrangement of the men in the wedges and in the opening plays from the center of the field, the formation is given which in the majority of cases would seem to be most advantageous. But this arrangement need not be considered fixed and may be changed at the discretion of the captain.

For special reason, too, it may in some instances seem best to alter the arrangement of the interference so that the positions of the preceding and following runners shall be interchanged. When there is sufficient reason for doing so, there should be no hesitation in making the alteration. When nothing is said as to duties of a player in the description of the diagrams, it will be understood that the player blocks his man.

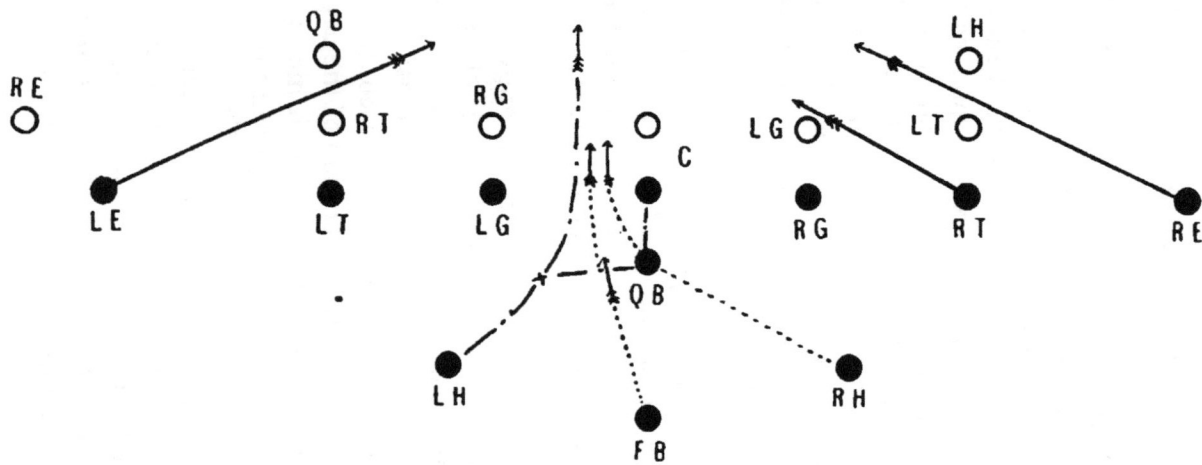

Diag 1

1. Half-back between guard and center on his own side.

To send LH between LG and C, the half-backs stand about *three** yards behind the rush line, directly in the rear of the opening between guard and tackle, FB stands directly behind center, about *four* yards from line, and the ends play *in* the line about *one and one-half* yards from tackles,* or as shown in diagram 5.

The *instant* the ball is snapped, LH, FB, and RH *dash* forward for the opening between LG and C; LH receives the ball from QB as he passes him on the run and strikes the line at *utmost speed* between LG and C, with *head down as low as the waist*, and the ball clasped tightly into his *stomach* with *both* arms. At the same instant the ball is snapped, LG lifts his man *back* and to the *left*, C lifts his man *back* and to the *right* to make an opening, while the ends and RT pass through the line at *full speed*, in the lines indicated, to be ahead of and interfere for LH in case he succeeds in getting through. FB and RH following directly behind LH at full speed, push him *with all their might* as he strikes the line. The instant QB has passed the ball he follows behind LH and helps push him.

NOTE. Many times when the runner is apparently blocked in the line he may be torn loose and carried on for long gains if all *plunge* and *tear* and *push* till the ball is "down." *Never* let *any* man cease work until "down" is called.

---

*The positions of the backs behind the line may vary from 2 to 4 yards, dependent upon the quickness of the men in starting.

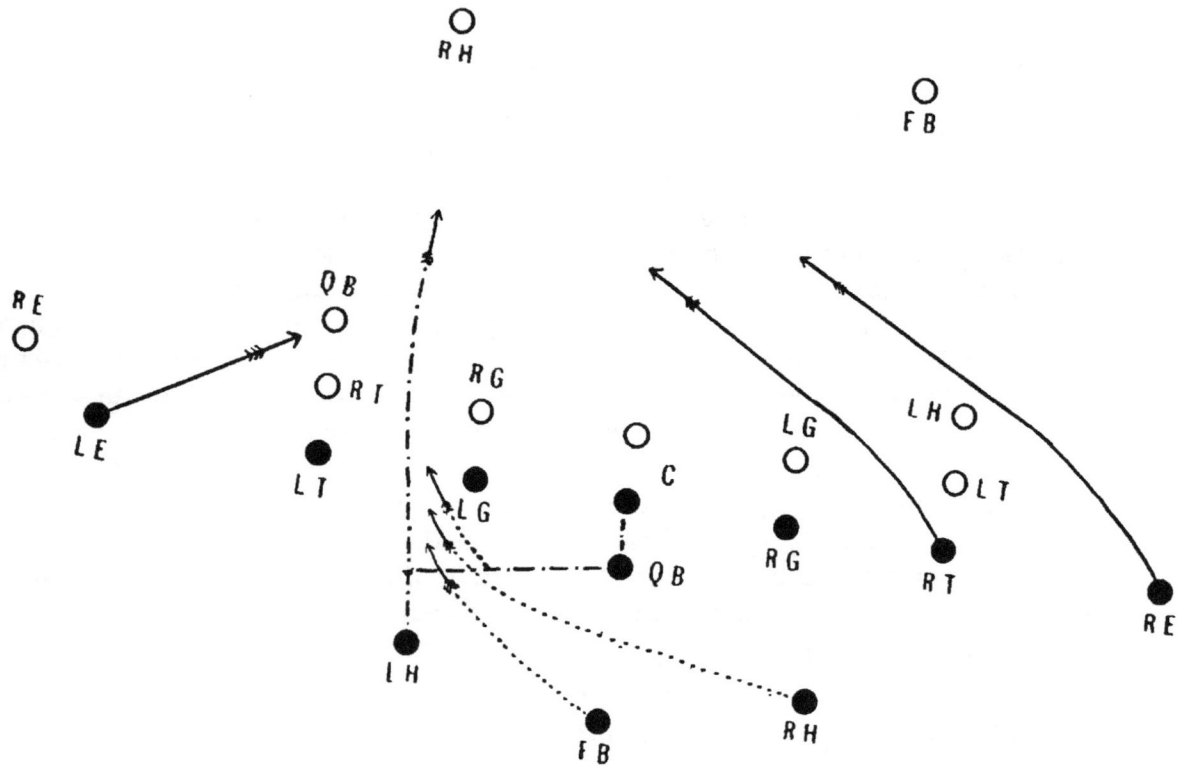

2. Half-back between guard and tackle on his own side.

To send LH between LG and LT, the backs and ends occupy *exactly* the same position as in play No. 1.

The *instant* the ball is put in play, LH, FB, and RH dash forward as before ; LH receives the ball at about x on a short pass from QB, and with *head down* and ball clasped at the stomach with both hands,* dashes into the opening between LT and LG, while FB, RH, and QB follow *directly behind* and push *with all their might* as he strikes the line.

LT lifts his man *back* and to the *left*, while LG lifts his man *back* and to the *right* the moment the ball is snapped, in order to open the line.

LE, RT, and RE also start the instant the ball is put in play ; LE dashes into the first man behind the opposing line, making sure at the same time that no one reaches LH from outside of LT before he strikes the line, while RE and RT take the directions indicated in the diagram, to arrive ahead of and interfere for LH as they go together down the field.

NOTE. It will the duty of RE and RT to block the opposing RH and FB, and each should make for the point in front of LH where he can best interfere with and block his particular man.

---

* It will be a great advantage upon emerging from the line to shift the ball to one arm, in order to have the other to use in warding off.

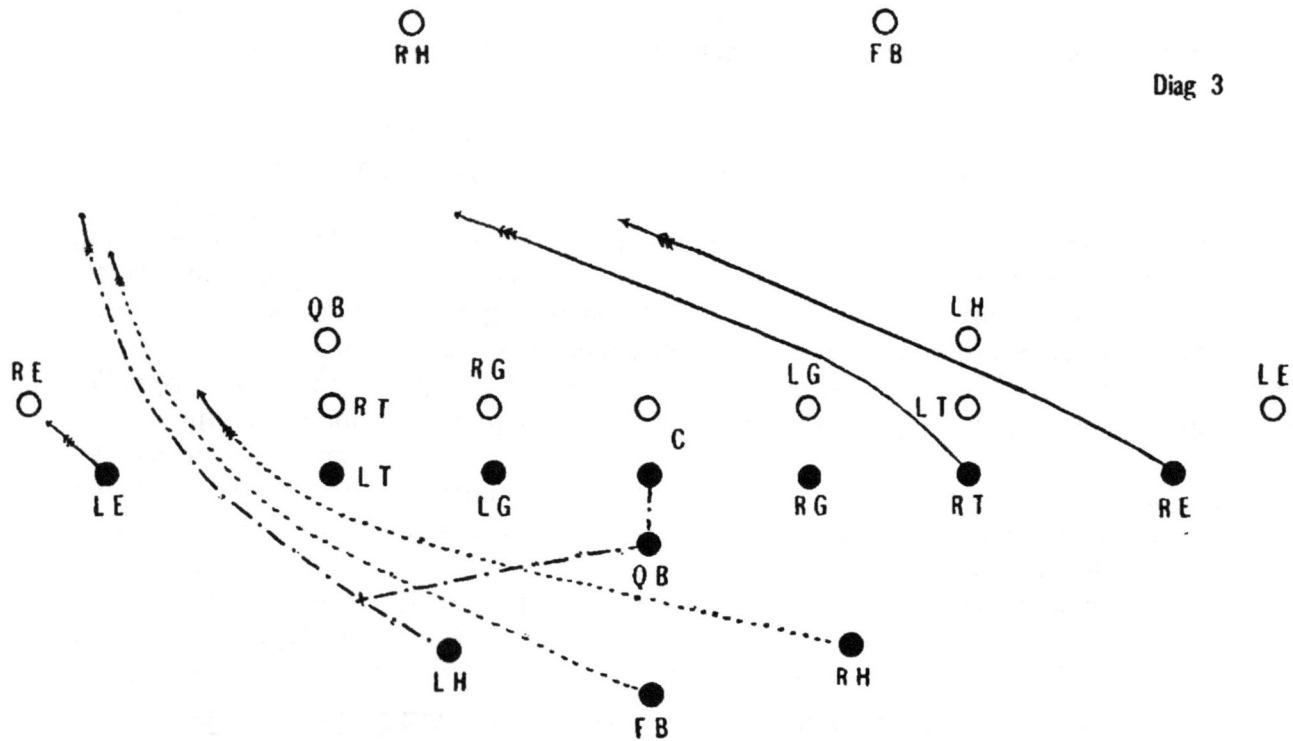

Diag 3

3.* Half-back between tackle and end on his own side.

To send LH between LT and LE, the backs and ends occupy the same position as in the preceding plays.

LH, FB, and RH start forward the *instant* the ball is snapped, as before, and LH receiving the ball at X on a pass from QB, dashes for the opening just to the *left* of LT, with his head down.

FB, and RH and QB follow directly behind, as in the preceding diagram, to throw their whole weight against LH when he strikes the line and push him through, in case he is blocked.

LT makes a supreme effort to carry his man *back* and in the *right*, while LE runs directly for the opposing end and endeavors to force him *out* to the left.

RT and RE, without stopping an instant to block their opponents, pass directly through the line and take the directions indicated in the diagram, to arrive in front of LH† at the left end and interfere for him in case he passes the line successfully.

---

* This play was made in the early stage of the development of the game, when the runner's ability to dodge was trusted to in order to make the play successful, but is now seldom if ever used.

† See NOTE, diagram 2.

Diag 4

4. Half-back around his own end.

To send LH around LE all the men occupy the same position as in the preceding plays of the series,* with the exception of LH, who shifts his position several yards to the left without attracting attention.

As before, LH, FB, and RH start forward at utmost speed the instant the ball is snapped, and LH, receiving the ball as he runs on a long pass from QB, sprints for the left side of the field in order to circle around and pass to the *outside* of the opposing end.

LE makes directly for his opponent, and endeavors to force him in toward the center, while FB, RH, and QB follow at utmost speed as before. FB and QB seek to overtake LH, running to the *inside* of him and interfering for him as they go together down the field, while RG follows as closely as possible behind LH to prevent his being caught from the rear.

RT and RE pass directly through the line and take the directions indicated as before,† to interfere for LH if he succeeds in circling the end.

---

*See explanation of diagram 1.
†See NOTE, diagram 2.

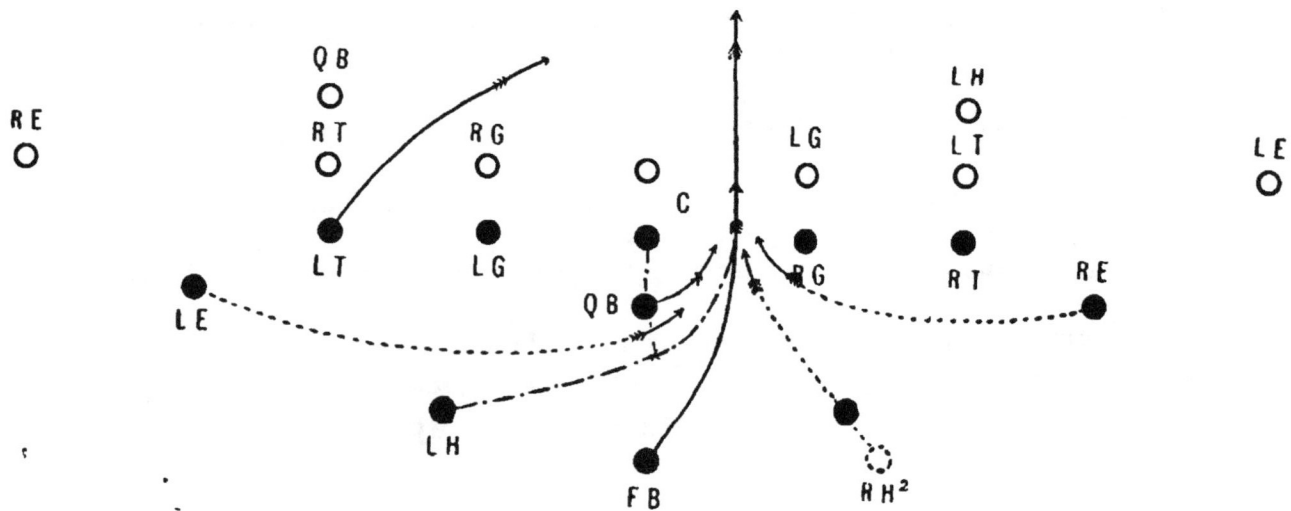

Diag 5

5. Half-back between guard and center on the opposite side.

To send LH between RG and C, the ends stand about *one* yard *back from the line* and a *yard* and a *half outside* of the tackles, the half-backs stand between two and three yards directly behind the *guards*, RH withdrawing slightly to RH², and the full-back stands between three and four yards behind the center.

The *instant* the ball is snapped FB, LH, RH, RE, and LE dash forward for the point between RG and C; RG lifts his man *back* and to the *right*, while C forces his man *back* and to the *left* to make an opening.

FB dashes straight into this space, passes directly through the line, breaking an opening, and jumps into the first man in his path behind the opposing line.

LH receives the ball from QB's hands as he passes on the run or by a short pass, and plunges into the opening directly behind FB with his *head down* and the ball tightly clasped at his *stomach* with *both hands.**

RH, RE, LE, and QB dash in immediately *after* LH, throw their *entire* weight against him and push him through.†

LT, simply forcing his opponent to pass *outside* of him, dashes in the direction indicated the instant the ball is in play, to arrive ahead of and interfere for LH in case he succeeds in getting through the line. It may be best for LT to select a *particular* back, and make it his especial duty to take *him* each time.

NOTE. RH may go in advance of LH together with FB, if so desired.

---

* See NOTE, diagram 2.
† See NOTE, diagram 1.

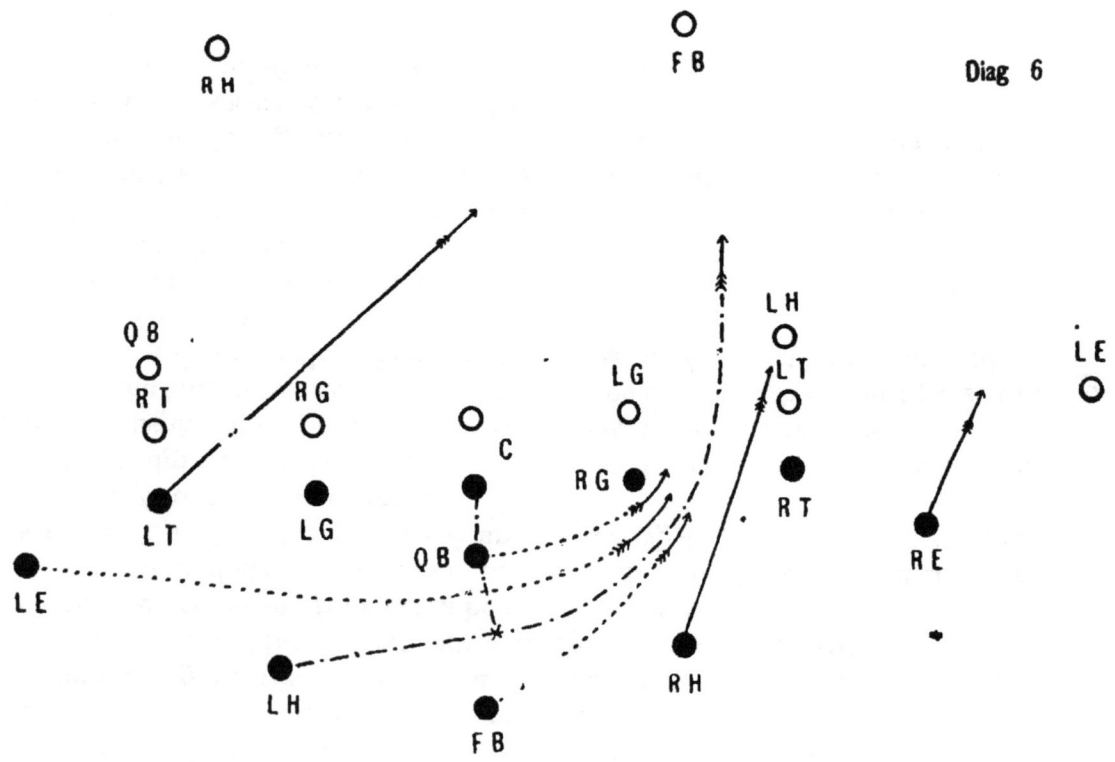

Diag 6

6. Half-back between the guard and tackle on the opposite side.

To send LH between RG and RT the men occupy the same position as in the preceding diagram.*

The *instant* the ball is snapped RH, FB, LH, and LE start forward at utmost speed in direction of lines indicated. RG lifts his man *back* and to the *left*, while RT lifts his man *back* and to the *right*.

RH dashes straight through the opening and takes the extra man behind the *opposing* LT. LH follows immediately behind and dives into the opening so made with head down, the ball held as before.†

LE leaves his position the moment the ball is put in play and follows directly behind LH.

FB and QB also dash in and all throw their combined weight in behind him as he strikes the line, to push him through.

The play of LT is the same as in diagram 5.

RE takes his own man and endeavors to force him *out* toward the side.

NOTE. FB may be sent through the opening with RH *ahead* of LH, to break the line and interfere, instead of following and pushing from behind.

* See description of positions of diagram 5.
† See description in diagram 5.

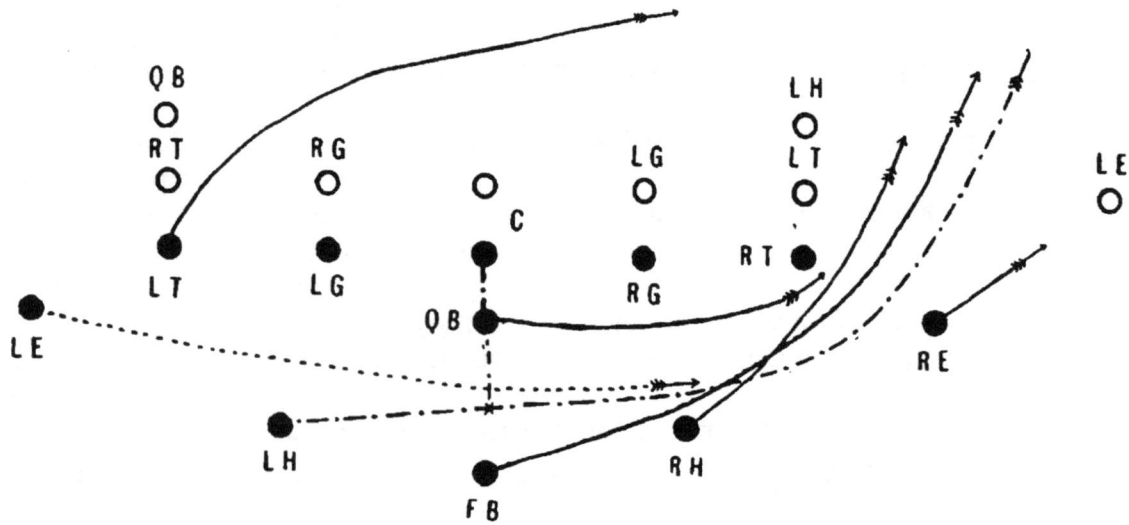

Diag 7

7. Half-back between tackle and end on the opposite side.

To send LH between RT and RE, the men take the same position as in the preceding play.*

As before, LH, FB, RH, and LE start in the direction indicated at *utmost speed* the instant the ball is snapped. RE takes his opposing man and forces him *out*. RH and FB dash for the opening to the right of RT ahead of LH, take the first men they meet after passing the line, and run in the direction indicated down the field to interfere for LH.

LH receives the ball on a pass from QB at X, makes for the opening at utmost speed, with *head up*, and as he turns down the field takes a line a little to the *outside* of RH and FB to have the benefit of their protecting interference.

QB should, if possible, seek to arrive at the opening *ahead of* and interfere for LH.

LE follows LH closely, to prevent him from being caught from behind.

LT, going through the line as before,* makes for the right side in the line indicated, to block the opposing backs.

NOTE. Care must be taken by RH and FB that they do not run so far ahead of LH as to diminish the value of their interference. They should precede him from one to three yards.

---

* See description, diagram 5.

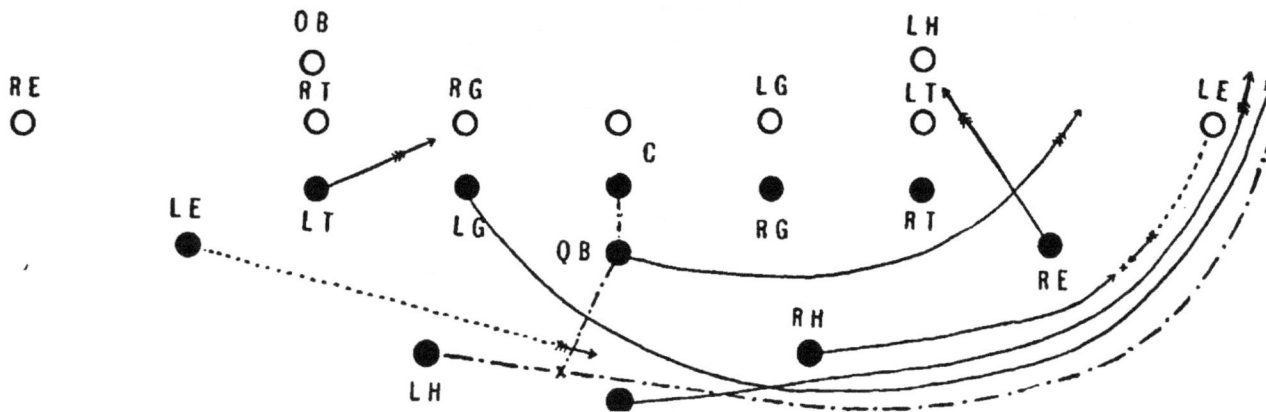

Diag 8

8. Half-back around the opposite end.

To send LH around RE there is no change in the position of the men behind the line.* As before, RH, FB, LH, LG, and LE start forward for the right end the *instant* the ball is snapped, at *utmost speed*. RT blocks his man and forces him as far as possible to the *left*. RE jumps directly into the line and either helps RT block his man, or takes the first extra man. RG blocks his man hard. RH runs straight for the opposing end-rusher, whom RE has left entirely exposed, meets him at about x, jumps into him and knocks him over or forces him in. FB following at RH's elbow will, if necessary, assist in blocking off the opposing end, and pass on down the field in the line indicated, to interfere for LH. LH receives the ball on a pass from QB on the run, and encircles the opposing end at top speed and passes down the field, a little to the *outside* of the line taken by LG.

LG breaks away from his opponent as the ball is snapped, and cutting in either directly behind QB or between QB and C, dashes for the right end, a little ahead of LH, and between him and the line, in order to interfere for him. A slow and lumbering guard may not attempt this play. LH may be obliged to withhold his speed until nearly at the end, in order to allow LG to get ahead of him.

QB *must* succeed in arriving at the end before LH. LE and LT play as in the preceding diagram,† or in case LG runs, LT leaves his own man to be taken care of by LE, and blocks the guard whom LG has left. This play requires the perfection of co-operation at every point, and can only be made successfully with constant practice. The attempt to have the guard run should not be abandoned because of numerous failures.

NOTE. In case LE, on the opposing side, plays far out, RH may force him still farther out, and FB and LH pass *inside* of him. Judgment must determine each time whether to pass the end on the *inside* or *outside*.

---

* See description, diagram 5.   † See description, diagram 7.

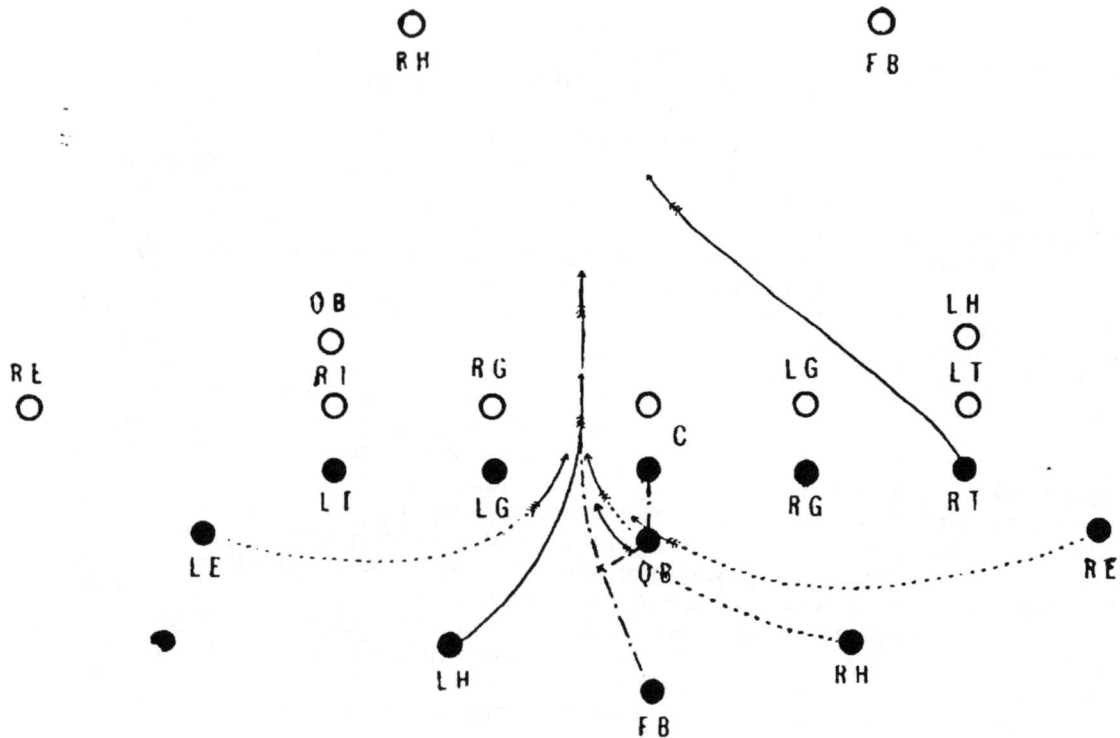

Diag 9

9. Full-back through the line between center and guard.

To send FB through the line, between LG and C, the men are placed as in the second series.*

The *instant* the ball is snapped, LH, FB, RH, LE, and RE dash forward for the opening between LG and C. At the same moment LG lifts his man *back* and to the *left*, while C carries his man *back* and to the *right* to widen the breach. LH rushes straight through the opening and down the field, making for the nearest back who opposes. FB, receiving the ball from QB as he passes on the run, plunges in directly behind LH, with his *head down*, and the ball clasped at his stomach with *both hands*.

LE, RH, RE, and QB rush in behind FB and throw their entire weight against him as he strikes the line, to push him through in case he meets with any resistance.†

RT slips through the line to the *inside* of the opposing tackle, without attempting to block him an instant, and takes the direction of the line indicated, to arrive ahead of and interfere for FB, in case he succeeds in passing the line.

LT and RG block their men.

NOTE. It may be well for RT to run directly for the opposing RH, and make sure that he is thoroughly blocked.

---

*See description, diagram 5.
† See NOTE, diagram 1.

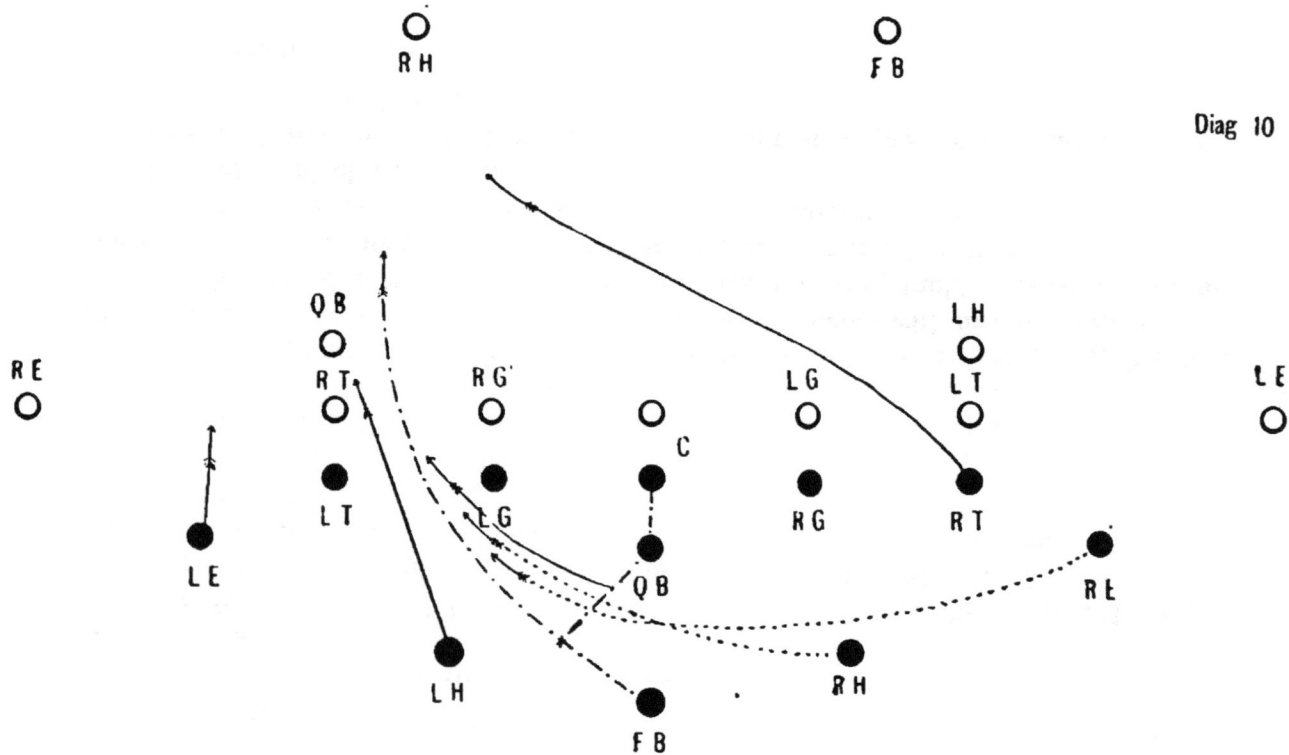

Diag 10

10. Full-back between the guard and tackle.

To send the full-back between LG and LT, the men stand as before.* All three backs and RE dash for the point between LG and LT, the instant the ball is put in play. LT lifts his man *back* and to the *left*, while LG lifts his opponent *back* and to the *right*. LH rushes through the opening ahead and takes the extra man behind the tackle.

FB receives the ball at x, on a pass from QB, as he runs, and dashes into the opening, directly behind LH, with his head down and the ball held as before.

RH, QB, and RE follow immediately behind FB, and throw their entire weight in to help him as he strikes the line.

LE takes the first man in the line outside of the tackle, and prevents the opposing end from coming in.

RT, without holding his man an instant, plays as shown in the preceding diagram.†

---

* See description, diagram 5.
† See NOTE, diagram 9.

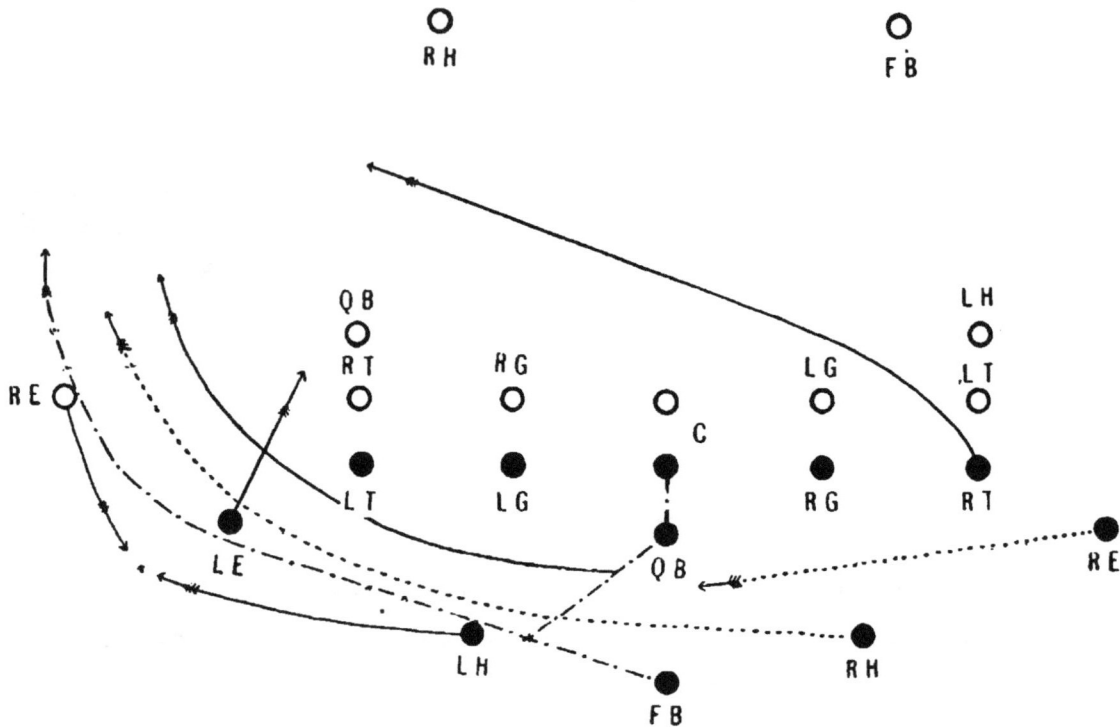

Diag 11

II. Full-back between the tackle and end.

To send the full-back between LT and LE, there is no change in the position taken by the men.*

The three backs and the ends dash forward in the lines indicated the instant the ball is snapped.

LE takes the tackle, the extra man behind the tackle, or the first man in the line outside of him.

LH runs directly for the opposing RE who has been left free, blocks him and endeavors to force him *out*.

FB receives the ball at x on a pass from QB and runs in the line indicated, at utmost speed, with head well up.

QB cuts in close behind LT and endeavors to get *ahead* of FB in order to give him interference.

RH and RE, running behind FB at utmost speed, seek to protect and assist him.

RT plays as before.†

LT exerts every power to force the opposing tackle *back* and in toward the center.

NOTE. The ball should always be carried in the arm *away* from approaching tacklers, both as a protection, and that the free arm may be that nearest the opponent, to be of use in warding off. It is fine play to shift the ball from arm to arm as occasion requires.

*See description of diagram 5.   †See diagram 8 and NOTE.

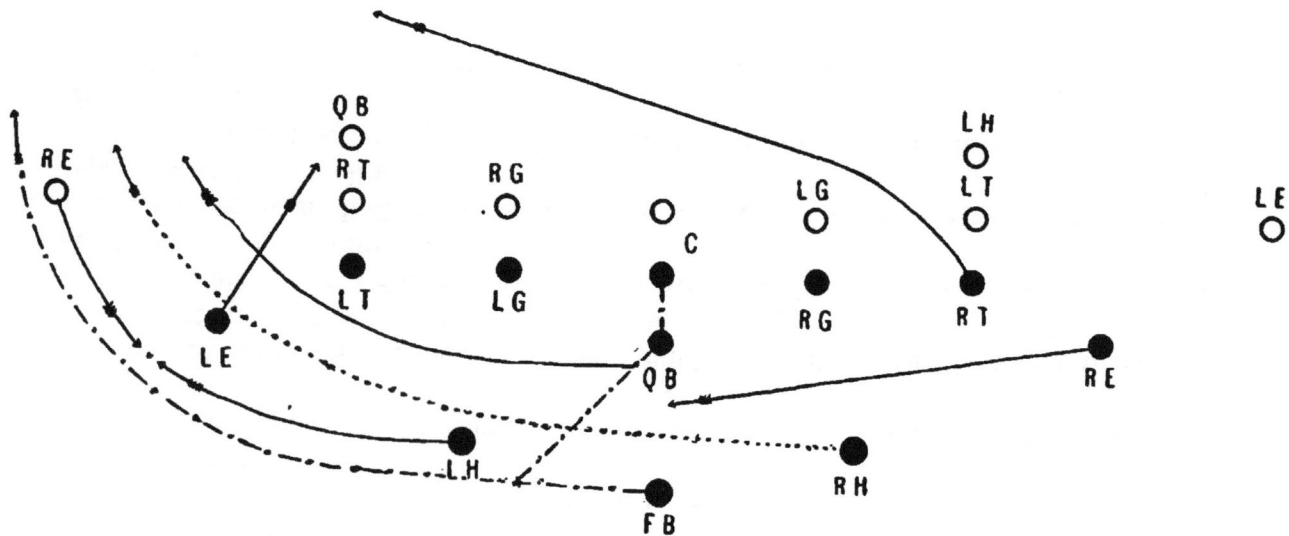

Diag 12

12. Full-back around the end.

To send the full-back around the end there is no change in the position of the men.

The play is made in almost identically the same way as shown in the preceding diagram, except that in the present case LH endeavors to force the opposing end toward the *inside*, while FB puts on utmost speed and rounds the end *outside* of him.

In all plays around the end circumstances may arise which offer an advantage in selecting the side of the opposing end other than that called for by the signal. While as a general rule it is best to follow the signal, for the interferers are working with that in mind, still, a skillful runner may secure long gains for his side by judiciously seizing an unexpected opening.

LE takes the first extra man in the line outside of tackle as before. (See diagram eleven.) Another play in which FB runs around the end is shown in diagram 63.

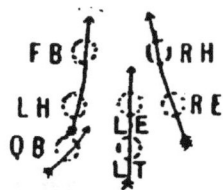

B Diag 13

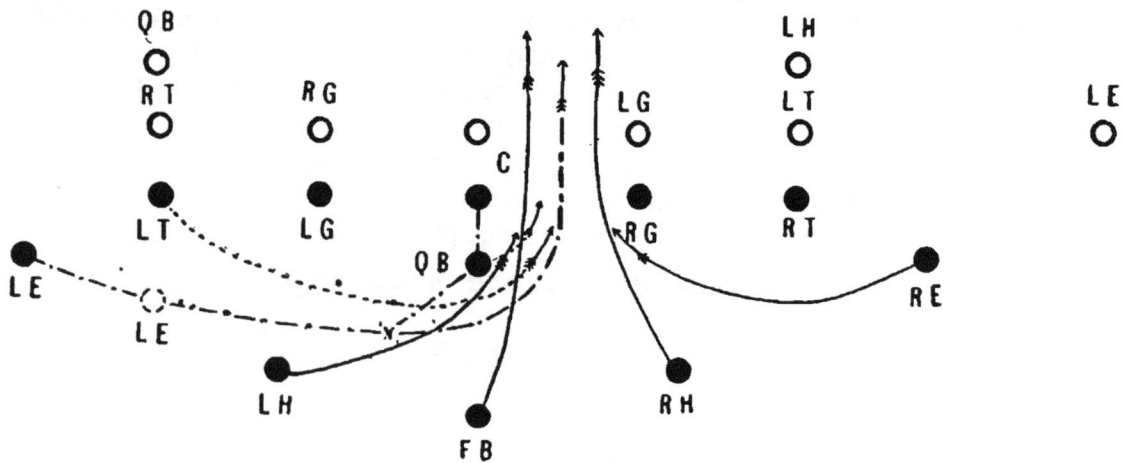

13. End between the center and opposite guard.

To send the LE between RG and C, the positions are the same as in the preceding series.*

The instant the ball is snapped the three backs and the ends dash forward for the point between RG and C, in the lines indicated.

C lifts his man *back* and to the *left*, while RG endeavors to force his man *back* and to the *right*.

FB and RH plunge through the opening abreast, and close together. LH follows directly behind FB and throws in his weight *as he strikes the line*, while RH is followed by RE in the *same manner*.

LE works in slightly to LE² before the ball is snapped and receives the ball from the hands of QB as he passes him; LE then turns in immediately behind and between LH and RE, carrying the ball in the same manner as shown for FB in play No. 1, diagram nine. A flying wedge is thus formed as the men strike the line at the point between C and RG. (See cut B.) QB falls in immediately behind LH and LE, while LT, who leaves his man almost instantly, follows directly in the rear of RE and pushes forward as the wedge strikes the line. (See cut B.)

NOTE. A vital point in the play is that LE be *close in behind* his interferers, and that the wedge, preserving its form as far as possible, strike the line with dash and

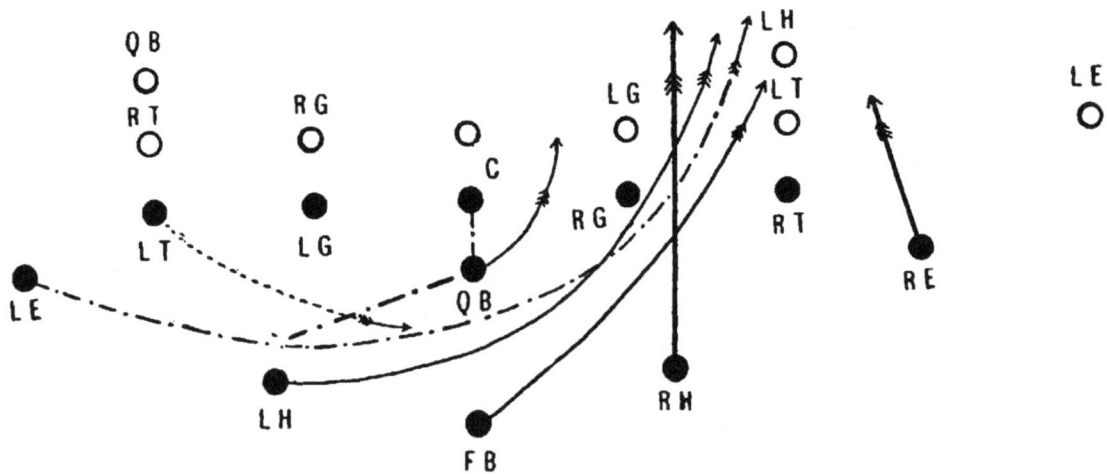

Diag 14

14. End between the opposite guard and tackle.

To send LE between RG and RT, there is no change in the position of the men.

The instant the ball is snapped RH, FB, LH, and LE dash forward for the point between RG and RT. RG lifts his man *back* and to the *left*, while RT forces his man *back* and to the *right*.

RH passes directly through close to RG, and, butting the opposing guard with his shoulder if he blocks the way, proceeds down the field and interferes with the first man that he meets behind the line. FB rushes diagonally through and runs directly into the opposing LT or the extra man behind the line, while LH dashes straight through the center of the opening and rushes down the field to interfere.

LE receives the ball at x on a pass from QB, and with *head down* dives into the line *directly behind* LH.

LT plays as shown in diagram thirteen.

After making the pass the best play for QB is to block the opening between RG and C, to prevent the opposing guard or center from coming through and getting LE before he strikes the line.

RE takes the first man outside the tackle, and prevents any one from passing around RT and stopping LE before he reaches the opening.

Diag 15

15. End between the opposite end and tackle.

To send the LE between RE and RT, preserve the same positions as in diagram fourteen.

RE plays as shown in diagram eight.

RH plays as LH in diagram eleven.

FB plays as shown in diagram seven.

QB plays as shown in diagram eight.

LH proceeds in the line indicated, at utmost speed, takes the first man on the opposing side as he rounds the tackle and continues on down the field to interfere. In case either the opposing LG or LT breaks-through the line QB must tackle him in order to prevent LE from being stopped before he reaches the end.

LT, leaving the line as shown in diagram thirteen, follows directly behind LE to make the play safe and prevent him from being overtaken from behind.

RT plays as shown in diagram eight.

LE receives the ball at x from QB, and, passing inside the opposing LE, turns down the field in the line indicated at utmost speed, passing to the *outside* of his interferers.*

NOTE.—The end must be careful to run just far enough behind the line to clear the opposing rushers as they break through.

---

* See NOTE, diagram seven.

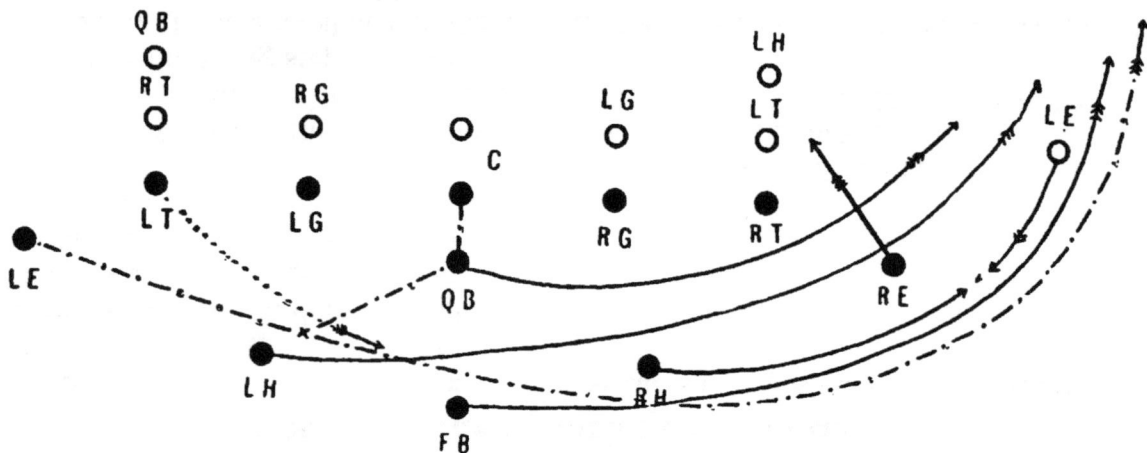

Diag 16

16. End around the opposite end.

To send the LE around the RE the play is made in identically the same manner as shown in diagram fifteen,* except that RH forces the opposing end-rusher *inside* instead of *out*, while FB and LE, after sprinting straight for the side line at utmost speed, turn down the field *outside* of the opposing end.† LG and LT might possibly play as shown in diagram eight, though there would be danger of having LE tackled from behind, as he would in this case have no protecting interferer behind and might be obliged to withhold his speed slightly in order to allow LG to get ahead of him.

NOTE.—This play, as nearly all end plays, depends for its success on the swiftness of the interferers and the man with the ball, and upon the *quickness* with which all *start*. Care must be taken that the interferers do not get too far in advance of the runner.‡

---

* See diagram fifteen.
† See NOTE, diagram eight.
‡ See NOTE, diagram seven.

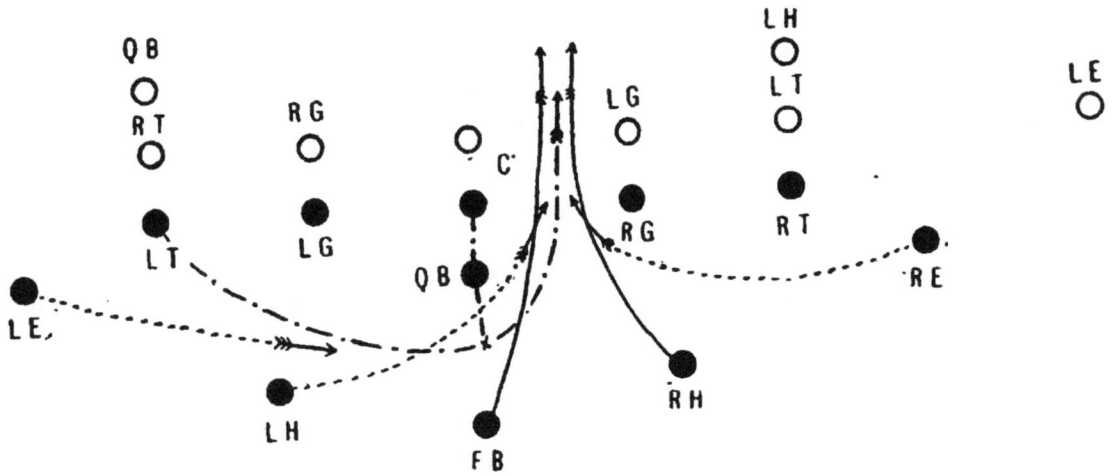

Diag 17

17. Tackle between the center and opposite guard.

To send the LT between C and RG, preserve the same position.

C and RG play as shown in diagram five.

RH and FB dash into the opening ahead of LT and take the first men they meet behind the opposing line.

LT breaks away from his man the instant the ball is snapped and receiving the ball from QB, turns in sharp around him as a pivot and plunges in behind FB and RH, with head down and the ball clasped at the stomach with both hands.

LH follows LT and plays as does LT in diagram ten.

RE rushes in behind LT and plays as shown in diagram five.

In case LT finds difficulty in getting away clear from the line, LE jumps in and takes the opposing RT as the ball is snapped, if necessary.

If LT is able to break away from his opponent without assistance, LE may follow directly behind LT to prevent his being caught from the rear and to push him through as he strikes the line.

QB follows LT, playing as shown in diagram five.

NOTE. LH may precede LT if it is thought best.

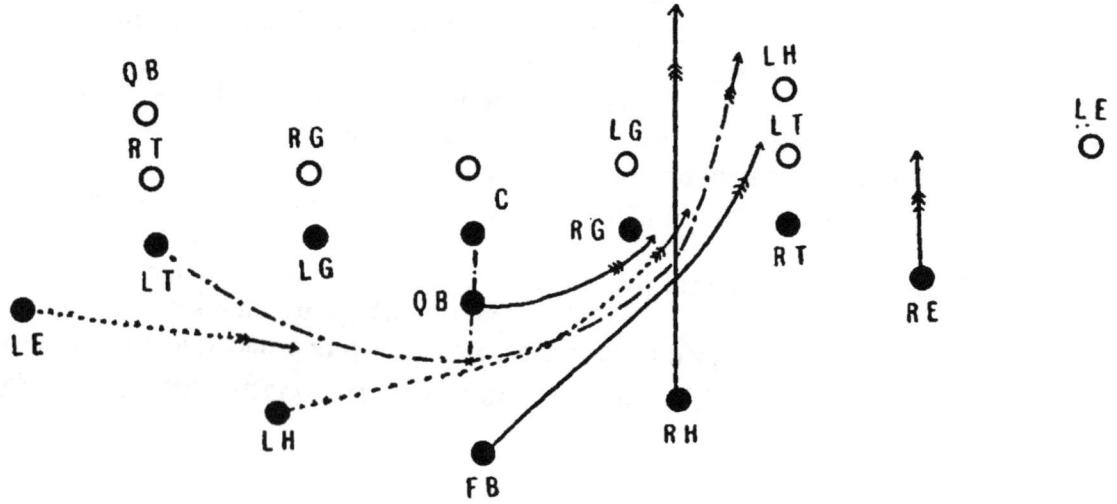

Diag 18

18. Tackle between the opposite guard and tackle.

To send LT between RG and RT, there is no change in position.

RE, RH, and FB plays as shown in diagram fourteen.

RH is nearer the opening and should pass through first. FB will cut in directly behind him, but both must take great care that they break *through* the line and are not stopped so that they choke up the opening, and are thus rendered of greater hindrance than help to the runner.

LT leaves the line as shown in the preceding diagram and dashes into the opening between RH and FB, with *head down* and the ball tightly held under the *right* * arm, or clasped at the stomach with both hands.

RG and RT play as shown in diagram six.

QB and LE following LT immediately, and push as shown in diagram seventeen.

LH also follows † directly behind LT to throw in his entire weight and push him through as he strikes the line, in case he meets with any resistance.

---

* When RT runs he will carry the ball in the left arm. In this way the ball will be kept on the side farther from the opponents where it will be less liable to be torn away, while it leaves the arm toward the opposing tacklers free for use in warding off.

† See NOTE, diagram seventeen.

Diag 19

19. Tackle between the opposite tackle and end.

To send the LT around the RT, the RH and FB all play as shown in diagram fifteen.

RE jumps directly into the line and either helps RT block his man, or takes the first extra man in the line.

LG and RG hold their men and force them back.

RT blocks his man and forces him as far as possible to the left.

QB plays as shown in diagram eight, or takes the opposing LT in case he succeeds in breaking through the line.

LT leaves the line as shown in diagram seventeen and carrying the ball in his *right* * arm plays as does LE in diagram fifteen.

LE jumps into the line and blocks the opposing tackle, if necessary, or follows LT and plays as shown in diagram seven.

LH follows close in the rear † of LT to prevent him from being tackled from behind and to assist him by interference as he rounds the end.

NOTE. In all plays around the right flank of the line, the rushers on that side must redouble their energy in order to make the play successful. When the play is on the left, the rushers upon that side will in turn block with their utmost power.

---

* See NOTE, diagram eighteen.

† See NOTE, diagram seventeen.

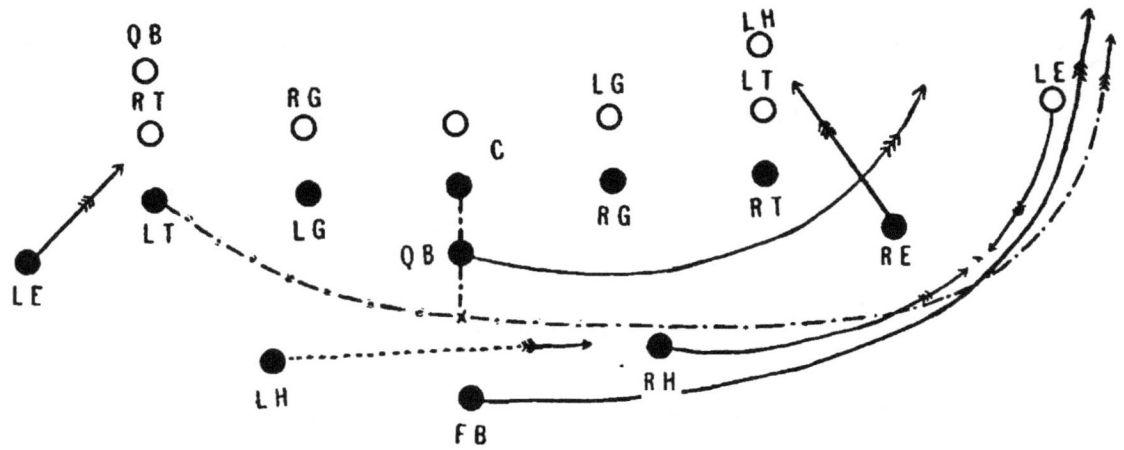

Diag 20

20.   Tackle around the opposite end.

To send LT around the right end, there is no change in the positions taken.

RE jumps into the line and helps RT block his man.

RH starts the moment the ball is snapped and runs directly for the opposing end who has been left exposed, bowls him over, or forces him in toward the line.

FB also starts with the snapping of the ball, and following almost directly behind RH passes slightly outside of him, helping block the opposing end-rusher if necessary, and then passes on ahead of LT to interfere for him in his run down the field.

RT plays as shown in diagram eight.

QB plays as shown in diagram nineteen.

LT leaves the line as shown in diagram seventeen, and taking the direction indicated, encircles the right end at utmost speed and plays as does LE in diagram sixteen.

LH and LE play as shown in diagram nineteen.

LG, C, and RG block their men.

NOTE.—It may be necessary in this play for LE to jump in and take LT's man, as he leaves the line ; otherwise he may follow and assist him as he rounds the end.

Diag. 21

21. Guard between opposite guard and center.

To send the LG between RG and C, the instant the ball is snapped LG jumps straight back from the line, breaking away from the opposing guard. He whirls directly around QB as a pivot and, receiving the ball from his hands as he passes, plunges into the opening between C and RG, with the ball held as shown in diagram one.

C and RG play as shown in diagram five.

RE, RH, FB, LH, and LE all start instantly and throw their entire weight in behind LG as he strikes the line, and force him through.*

QB also follows immediately behind LG and plays as in diagram five.

LT and RT block their men.

NOTE.—Instead of following behind LG it is often better that RH should draw slightly nearer the line before the ball is snapped, dash into the opening ahead of LG, and play as does FB in diagram five.

*See NOTE, diagram one.

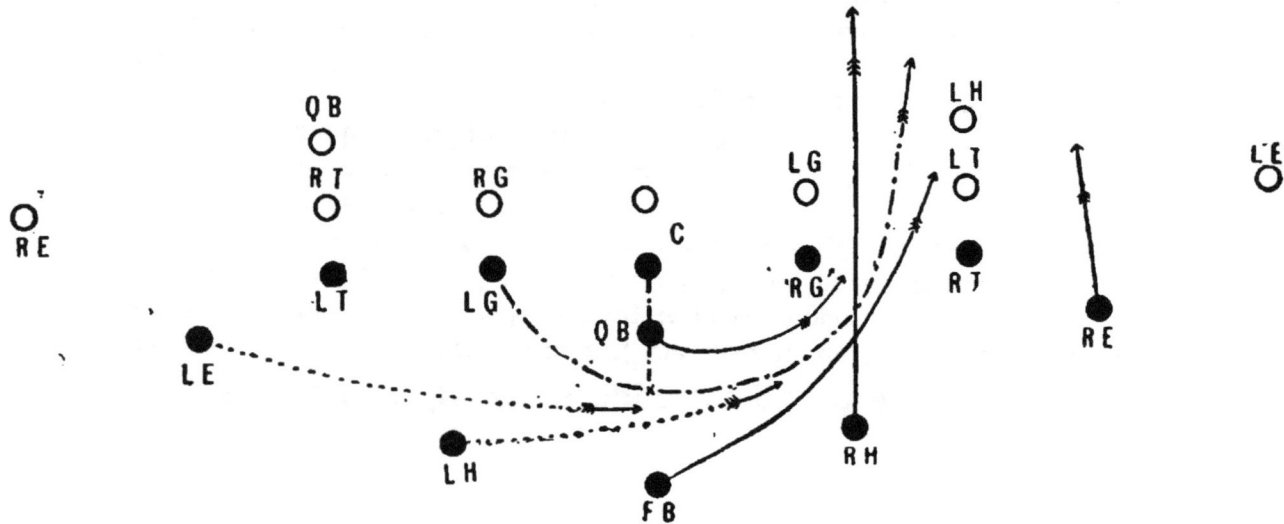

Diag. 22

22. Guard between the opposite guard and tackle.

To send the LG around between RG and RT, LG breaks away from his opponent the instant the ball is snapped, as shown in diagram twenty-one, receives the ball from QB as before, and dashes into the opening with head down.

RG and RT play as shown in diagram six.

RH starts forward the instant the ball is snapped and, dashing into the opening between RG and RT, strikes the opposing LG with his shoulder with the greatest possible force as he passes through, and then proceeds on and takes the first man behind the line.

FB crosses behind RH and rushing into the same opening plunges into the opposing tackle or the man immediately behind him.

RE plays as shown in diagram fourteen.

QB, LH, and LE follow behind LG and play as shown in diagram eighteen.

LT plays as in the preceding diagram.

NOTE.—RH and FB must see to it that they break *through* the line and are not there blocked so that they fill up the opening through which LG, who is following immediately behind, is to pass.

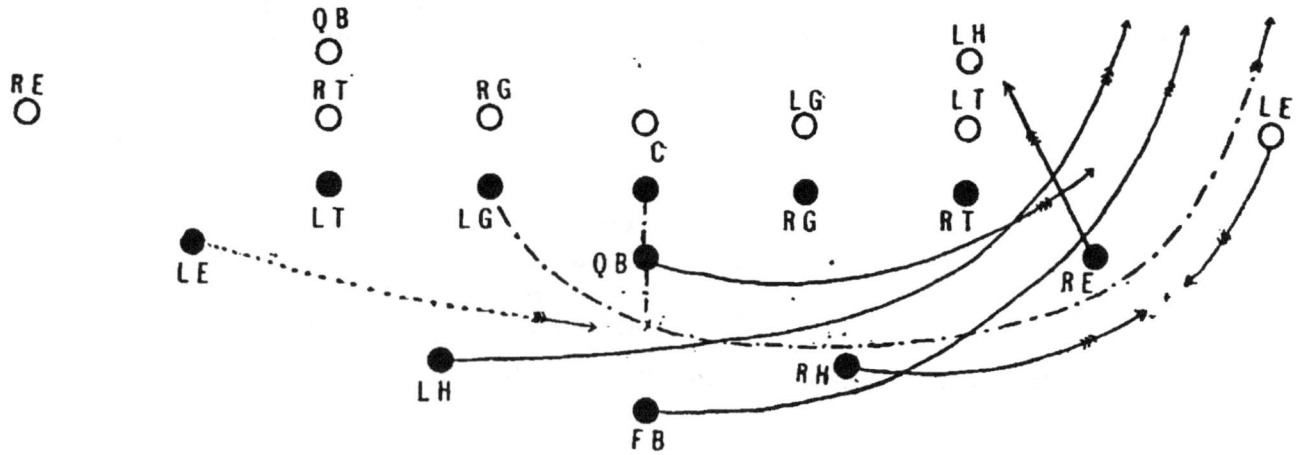

Diag. 23

23. Guard between the opposite tackle and end.

To send LG around the opposite tackle, LG leaves the line as shown in diagram twenty-one, and receiving the ball from QB, plays as does LE in diagram fifteen.

RH, FB, and LH start for the right end at utmost speed the instant the ball is put in play. RH runs directly for the opposing LE, forces him out or bowls him over.

FB and LH cut in, in the lines indicated, and passing outside of RT interfere by taking the first opponents whom they meet on arriving at the end.

RE and RT play as shown in diagram fifteen.

QB having passed the ball, runs at utmost speed by the *side* of LG if possible, *between* him and the line, in order to give him protection and assist him by interference.

LE follows directly in the rear of LG to make the play safe and prevent him from being overtaken from behind.*

LT and RG block their opponents.

---

* See NOTE, diagram seven.

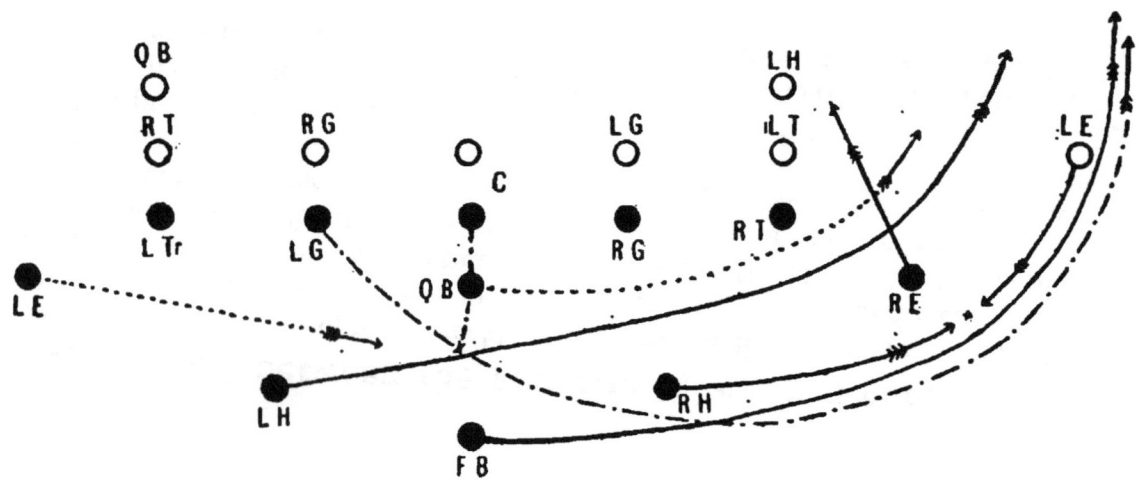

Diag. 24

24. Guard around the opposite end.

To send LG around the right end, LG leaves the line as shown in diagram twenty-one ; receives the ball at x from QB and passes around the right end at full speed, swing back several yards from the line.

RH, FB, and LH dash toward the right end the instant the ball is put in play, RH runs directly for the opposing end, who has been left exposed, and forces him in or blocks him off.

FB assists RH if necessary, and then passes on around the end ahead of LG to interfere.   LH crosses in front of LG and cutting in close behind RT blocks the first free man on the opposing side.

RE and RT play as shown in diagram twenty.

LT and QB play as in diagram twenty-three.

LE follows LG to protect him from behind and to make the play safe.

Diag. 25.

25. Criss-cross half-back play around the end.

To send LH around the right end on a criss-cross between the half-backs, the men stand in their regular positions with the exception of LH, who works out nearer LE and slightly back, without attracting attention, to the second position as shown in the cut.

The instant the ball is snapped, RH starts in the direction indicated, receiving the ball at x, and passing close in front of LH, carries the ball at his left side, so that LH may receive it from him as he rushes by, and proceeds on in the line indicated; LH *stands in his tracks* until RH nearly reaches him, and upon securing the ball instantly starts in the opposite direction at utmost speed and passes around the right end.

RT, RG, and RE play as shown in diagram eight, FB works slightly to the left before the ball is snapped, and *stands still* until RH nearly reaches LH, and starting forward as indicated at the same instant with LH, makes directly for the opposing LE, and blocks him or forces him *in*.

LG breaks away from his man the moment that LH receives the ball and plays as in diagram eight. QB having passed the ball, stands still until LH has received it, and then plays as in diagram eight, or in case an opponent comes through the line between C and RT, it is the duty of QB to attend to him. LT blocks his man hard or blocks the opposing RG, left exposed by LG. LE follows LH and protects him from behind as shown in diagram eight.

NOTE. In all end and criss-cross plays, great care should be taken that the runners do not pass so close to the line that their own men will be pushed back upon them, or so far in the rear that time and space will be lost.

6

Diag. 26

26. End criss-crosses with half-back in play around the end.

To send the RII around the left end on a pass play from LE to RII, LE draws in slightly from his original position to LE^2, and as the ball is snapped, dashes for the right end, receiving the ball at **x**, and carrying it at his right side, passes on close in *front* of RII, as indicated.

RII stands still in his position until LE nearly reaches him. On receiving the ball RII dashes in the opposite direction and passes around the left end at utmost speed.

As RII receives the ball, FB, LII, QB, and RG, who have, until this moment, stood in their positions, dart for the left end, in order to precede RII and interfere for him.*

LT and LG block their men with the *greatest energy*.

LII makes directly for and blocks the opposing RE.

FB takes the direction indicated, helping LII block his man, if necessary, and then proceeds on down the field.

QB plays as shown in the preceding diagram, while RG, breaking away from his man, plays as does LG in diagram twenty-five.

RT also plays as does LT in diagram twenty-five.

RE follows immediately behind RH to protect him from the rear.

NOTE. LII may precede LE and take the first man in the line outside of RT, to aid in the deception; in which case FB will take the opposing LE, while the others play as before.

---

*See NOTE, diagram twenty-five.

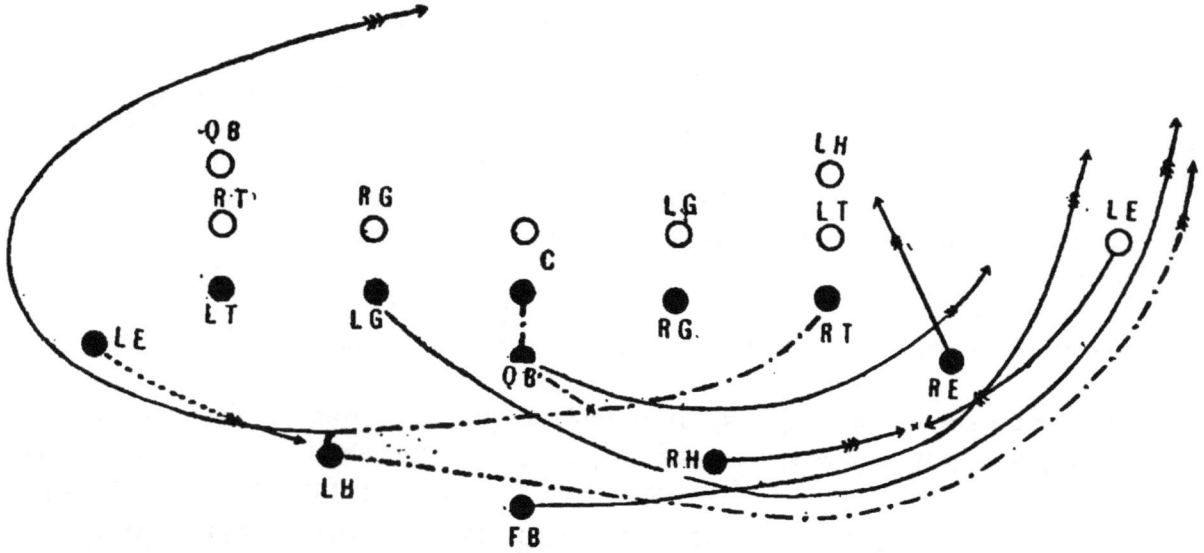

Diag. 27.

27. Tackle criss-crosses with half-back in play around the end.

To send the LH around the RE on a pass play from RT, RT leaves the line the instant the ball is snapped, and receiving it at x runs close in front of LH, passes him the ball as he goes by.

RE jumps in and takes the man left exposed by RT as he leaves the line. RG holds his man hard.

As LH receives the ball, QB, LG, LT, and LE all play exactly as shown in diagram twenty-five.

RH and FB stand still until the instant that LH receives the ball, and then play as shown in diagram eight.*

NOTE. To make the deception more complete, RH may start the instant the ball is snapped, and directly precede RT and take the first man he meets on the other side of LT ; in that case FB will take the opposing LE.

*See NOTE, diagram twenty-five.

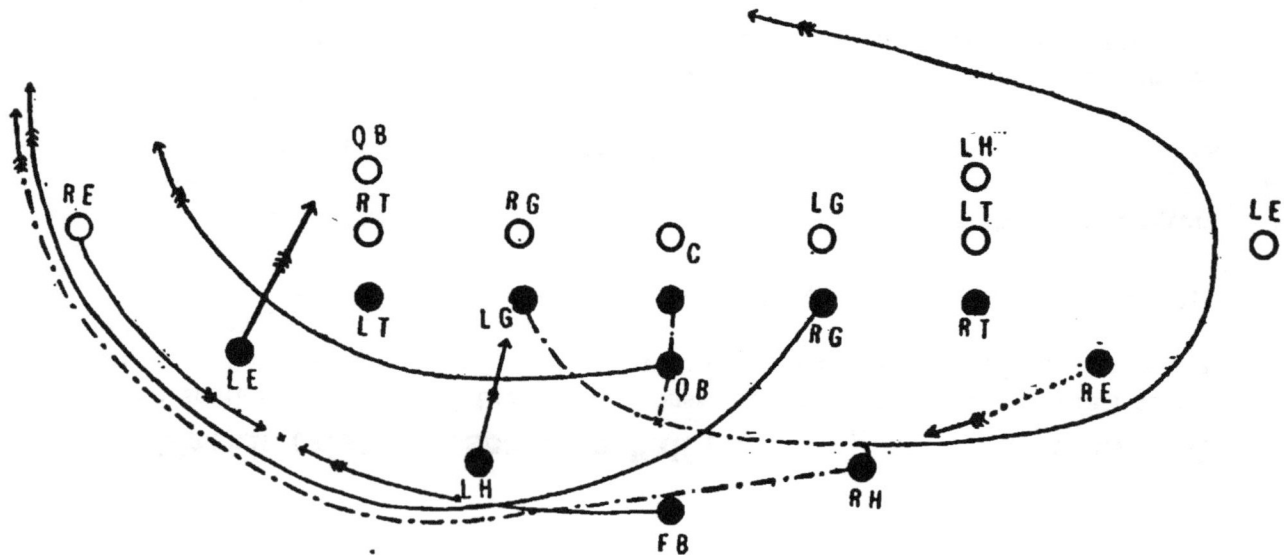

Diag. 28

28. Guard criss-crosses with half-back in play around the end.

To send the RH around the left end on a pass from LG, LG leaves the line as the ball is snapped, as shown in diagram twenty-one, and passes directly in *front* of RH, giving him the ball as he rushes by.

As LG leaves the line, LH dashes in and takes the opposing RG, who has been left exposed.

At the moment that RH receives the ball, FB, QB, and RG start for the left end, in the lines indicated, in advance of RH, to interfere for him.

FB runs directly for the opposing RE and bowls him over or forces him in.

RE, RT, RG, QB, and LT play as shown in diagram twenty-six.

LE helps LT block the opposing tackle or takes the inside man in the line.

Diag. 29

29. Half-back criss-crosses with the end in play around the opposite end.

To send LE around the RE on a pass from RH, LE draws slightly in and back to LE^2, nearly on a line with the half-backs, before the ball is snapped.

RH dashes forward as the ball is put in play, receives it at x, and passing close in *front* of LE gives him the ball, and then rushes directly into the opposing LE.

LE, having received the ball, starts for the right end in the line indicated, trusting altogether to his speed, as there is no one to protect him from behind.

RE may either jump into the line and help RT block his man, or take the first man that comes through on the right side of the rush line and force him in toward the center.

RT, RG, LG, LT, and FB play as shown in diagram twenty-five.

It may be necessary for QB to precede RH and take the first extra man in the line on the left end, in order to allow LE to get away with the ball without being caught from behind. Otherwise he will play as shown in diagram twenty-five.

In case a man comes through the rush line on the right of c, where QB is playing as shown in the cut, it is his duty to block him.

NOTE. FB may precede RH and play as LH in note on diagram twenty-six.

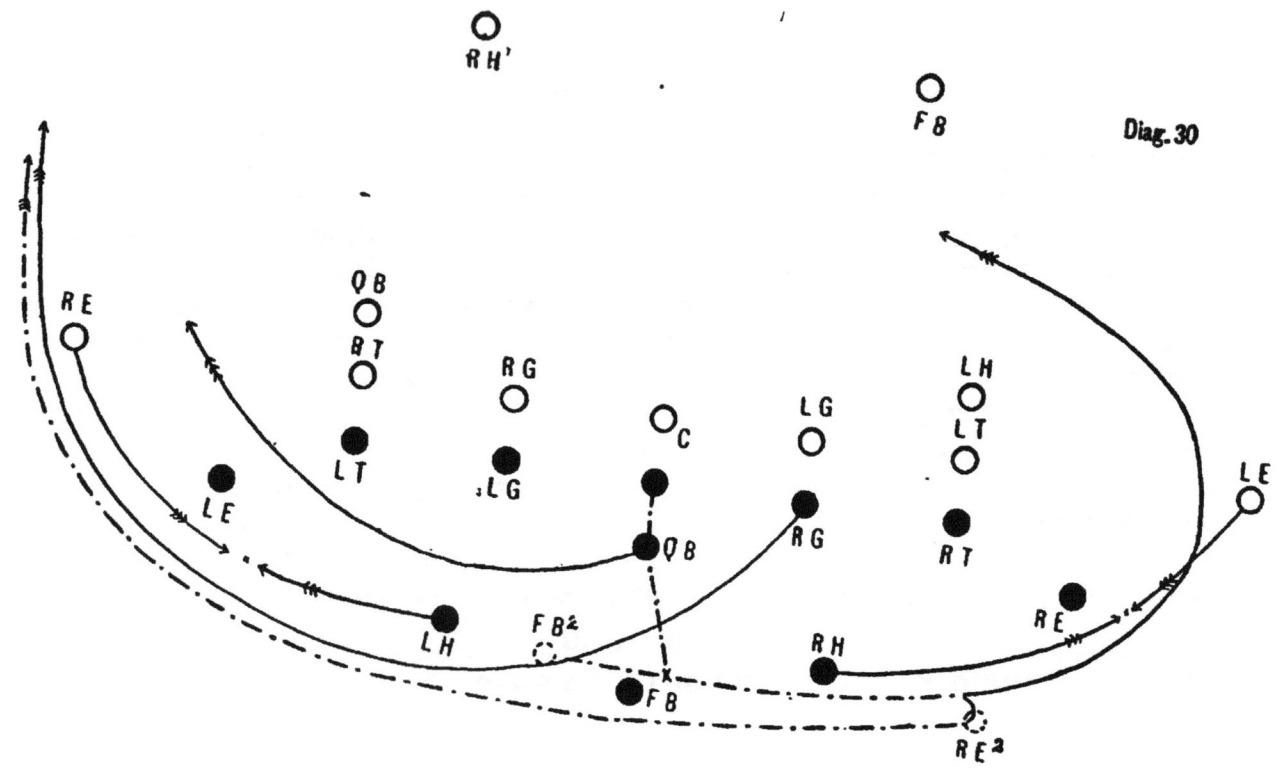

Diag. 30

30. Full-back criss-crosses with the end in play around the opposite end.

To send RE around the left end on a pass from FB, RE works slightly in and on a line with RH, at RE², before the ball is snapped, while FB moves a little to the left to FB².

As the ball is put in play RH and FB dash toward the right, FB receiving the ball on a pass at X.

RH rushes directly for the opposing LE, or takes the first extra man in the line, while FB runs close in *front* of RE to whom he gives the ball as he passes.

Upon receiving the ball RE instantly starts in the opposite direction and encircles the left end at utmost speed.

RG and LH precede RE and play as shown in diagram twenty-six.

RT plays as does LT in diagram twenty-five.

LE and LT play as in diagram twenty-eight.

LG blocks his man hard.

In case anyone succeeds in breaking through the line to the left of center, QB immediately blocks him. He will otherwise play as in diagram twenty-eight.

Diag. 31

31. Ends criss-cross and play around the end.

To send LE around the right end on a pass from RE, LE and RE both play in slightly, while LE works back until he is nearly on a line with the half-backs.

RE and RH start toward the left the instant the ball is snapped, RH preceding and taking the first extra man in the line beyond the tackle.

RE receives the ball at X on a pass from QB, and running close in *front* of LE, passes him the ball and rushes on into the opposing RE.

As RE reaches LE, FB, LH, and LG dash toward the right in the lines indicated.

FB takes the first extra man in the line beyond RT, LH runs directly for the opposing LE, and LG, preceding LE, plays as shown in diagram twenty-five.

QB blocks the first man through on the right hand side of the center, if necessary, or plays as shown in diagram twenty-five.

LE, upon receiving the ball, starts toward the right at utmost speed, keeping just to the outside of LG.

LT, RG, and RT play as shown in diagram twenty-five.

Diag. 32

32. Tackle criss-crosses with the end in play around the opposite end.

To send RE around the left end on a pass from LT, before the ball is snapped RE works in and slightly back to RE².

As the ball is put in play LT leaves the line as shown in diagram seventeen, receives the ball at x, runs close in front of RE, and gives him the ball as he passes. LT then blocks the opposing LE.

LH precedes LT and plays as does RH in diagram thirty-one.

FB plays as shown in diagram twenty-eight.

RG and RT play as shown in diagram twenty-six.

QB plays as shown in diagram thirty.

RE plays as does LE in diagram twenty-nine.*

As LT leaves the line, LE jumps in and takes the opposing tackle as in diagram seventeen.

LG plays as shown in diagram thirty.

RH starts forward in the line indicated as RE receives the ball, and precedes him around the left end, crossing in front of RG as the latter swings in behind the line.

---

*See NOTE, diagram twenty-five.

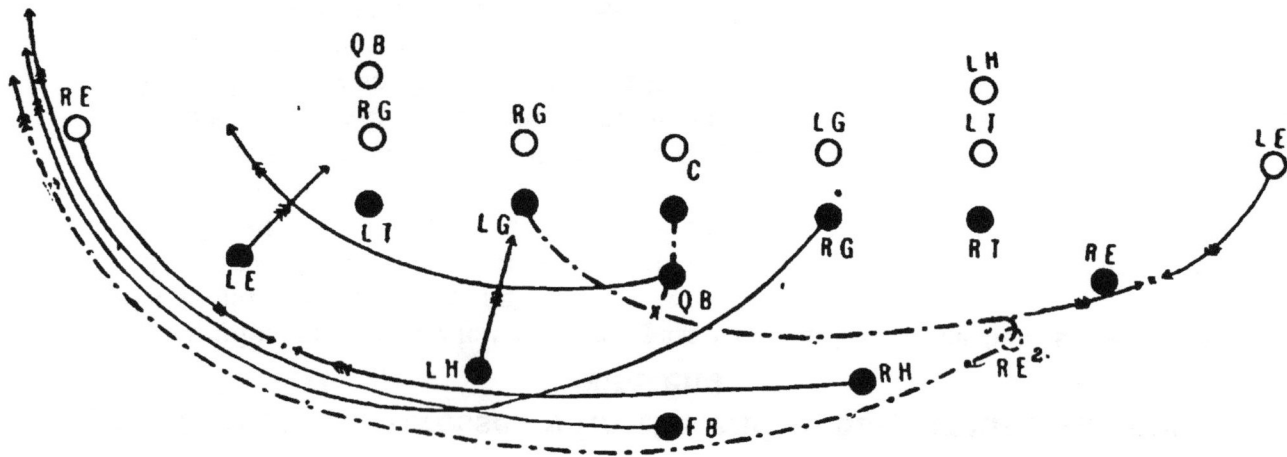

Diag. 33

33. Guard criss-crosses with the end in play around the opposite end.

To send RE around the left end on a pass from LG, RE draws in and slightly back to RE².

As the ball is snapped LG breaks away from his man as in diagram twenty-one, receives the ball at X, passes close in front of RE and rushes directly into the opposing LE.

RE remains in his position until LG has almost reached him, receives the ball from LG as he runs by, and plays as in the preceding diagram.

LH plays as shown in diagram twenty-eight.

QB, RH, and FB remain standing in their positions until the instant that RE receives the ball. They then dash toward the left and precede RE at greatest speed to interfere for him.

FB runs directly for the opposing RE. RH follows FB and assists him to block the end if necessary or continues on around the end and takes the first free opponent.

RG leaves the line as LG reaches RE, and plays as shown in diagram twenty-six.

All the other men play as shown in diagram thirty-two.

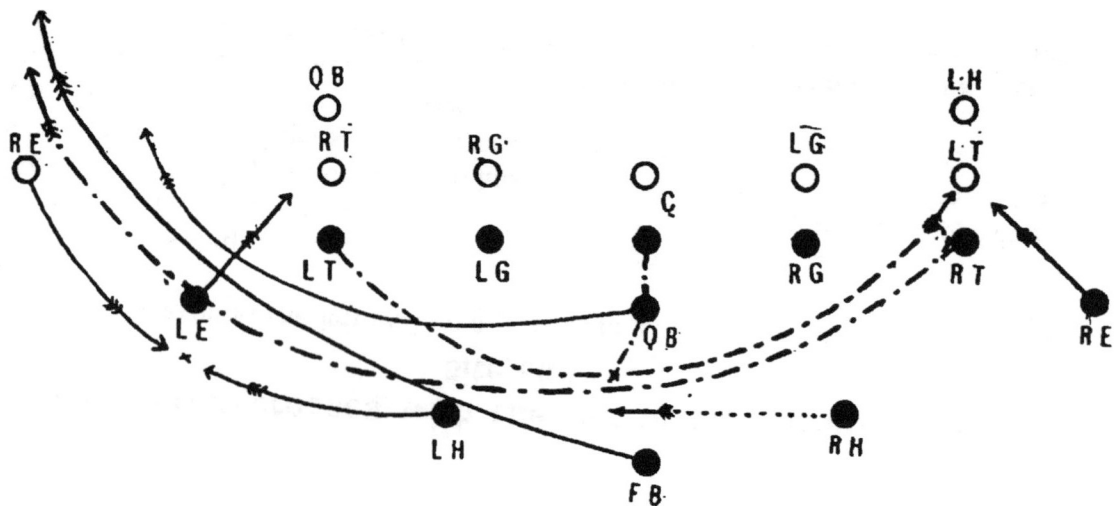

Diag. 34

34. Tackle criss-crosses with tackle in a play around the end.

To send RT around the left end on a pass from LT, the instant the ball is snapped LT plays as shown in diagram eighteen and runs directly for RT.

As LT reaches the line RT jumps suddenly back and receives the ball from LT as he passes, while LT rushes on directly into the arms of the *opposing* LT.

RE jumps in and helps block the opposing LT or takes the first extra man in the line.

LE blocks the opposing tackle as LT leaves the line.

LH, FB, and RH stand still until RT receives the ball. Upon receiving the ball RT instantly starts back in the opposite direction, taking the line indicated, and circles the left end with QB, LH, and FB in advance as interferers, and RH to follow and protect him from behind.

NOTE. RH may also run in advance of RT to interfere when the opposing LE has not come around the end of the line.

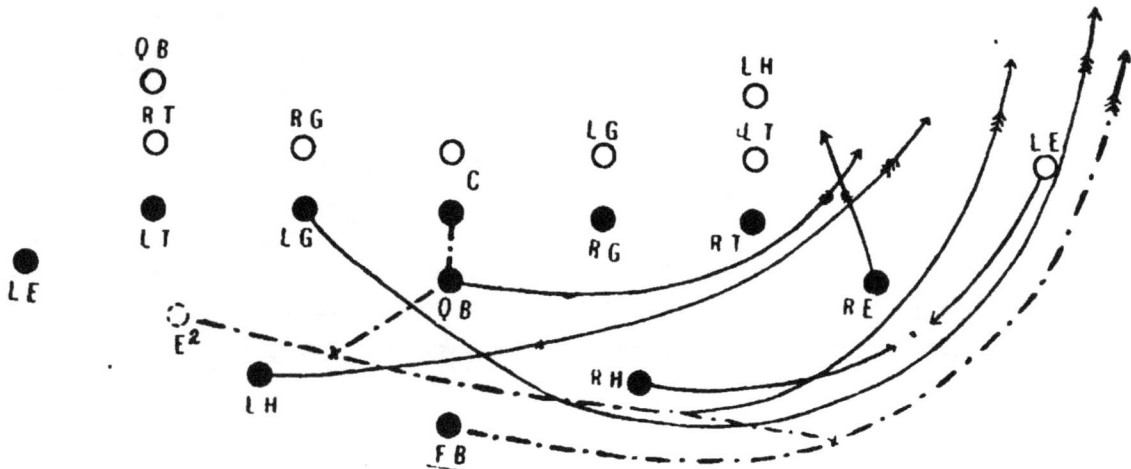

Diag. 35

35. Double pass from end to full-back in play around the end.

To send FB around the right end on a double pass from LE, LE should work in slightly before the ball is snapped to LE², as in diagram twenty-six.

The instant the ball is put in play, RH, FB, LH, LG, and LE leave their positions and dash for the right end, in the lines indicated in the diagram.

QB, RE, RT, LG, LT, and RG play as in diagram eight.

LH cuts in close around RT, and blocks the first man he meets.

RH runs directly for the opposing LE, and blocks him off or forces him in.

LE receives the ball at x, on a pass from QB, and running at full speed in a course, about two yards nearer the line* than that taken by FB, throws the ball on ahead of him to the full-back, with a clean pass of from four to five yards, as FB reaches the point behind RT, and turns sightly sidewise, as he runs, to receive it. LE then cuts in, as shown, and blocks the first free man on the opposing end.

FB, upon receiving the ball, passes to the outside and encircles the end.

The play may sometimes be made to greater advantage by having LH take the line indicated for FB, and receive the ball on the pass, while FB runs on ahead. The guard will find great difficulty in getting in advance of the man with the ball, and may find that he can be of more service by cutting across close behind the line.

NOTE. It may be necessary for FB to withhold his speed slightly until LE has passed him the ball. If it becomes necessary to block the opposing RG, LH may play as in diagram thirty-three, though LG should so time his action that his man will be unable to interfere with the play.

---

* See NOTE, diagram twenty-five.

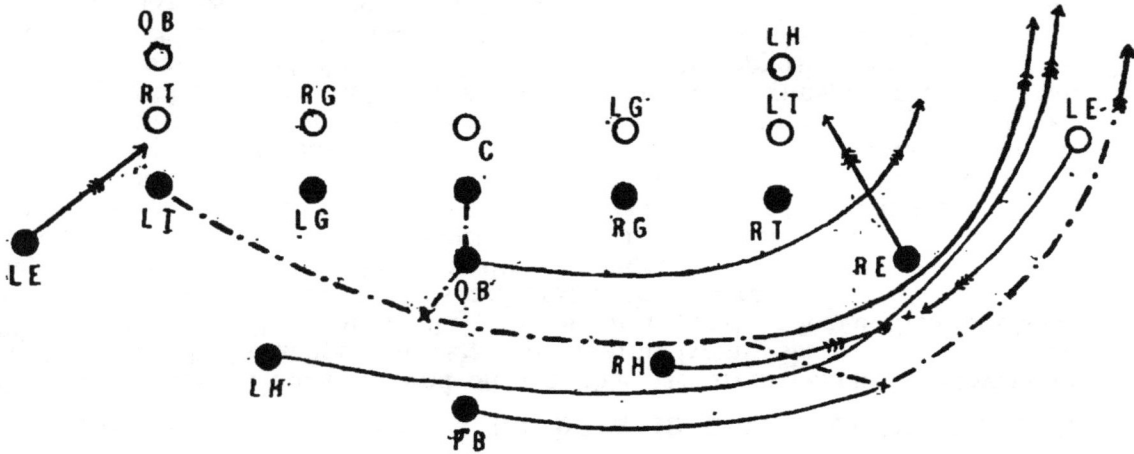

Diag. 36

36. Double pass from tackle to full-back in play around the end.

To send FB around the right end, on a double pass from LT, there is no change from the regular formation in the primary arrangement.

The instant the ball is snapped, LT leaves the line, receives the ball at x from QB and starts for the right end, precisely as shown in diagram nineteen.

LE jumps into the line and takes LT's man as he leaves him.

RE, RT, RG, and LG play as shown in diagram eight.

QB also plays as shown in diagram eight.

LH, FB, and RH, all start for the right end the moment the ball is snapped.

RH runs directly for the opposing LE, and bowls him over or forces him in.

LH assists RH, if necessary, and then cuts in down the field, as indicated, to interfere.

As FB is about to round the end, he turns half around without slackening speed, and receives the ball at about x, on a clean pass from LT. LT then turns in to interfere on the end, while FB passes on encircling the opposing LE.

NOTE. The pass may be made with equal success to LH; in which case FB will assist RH in blocking his man, and then pass on down the field to interfere, while LH swings out in a course just outside of the opposing end-rusher.

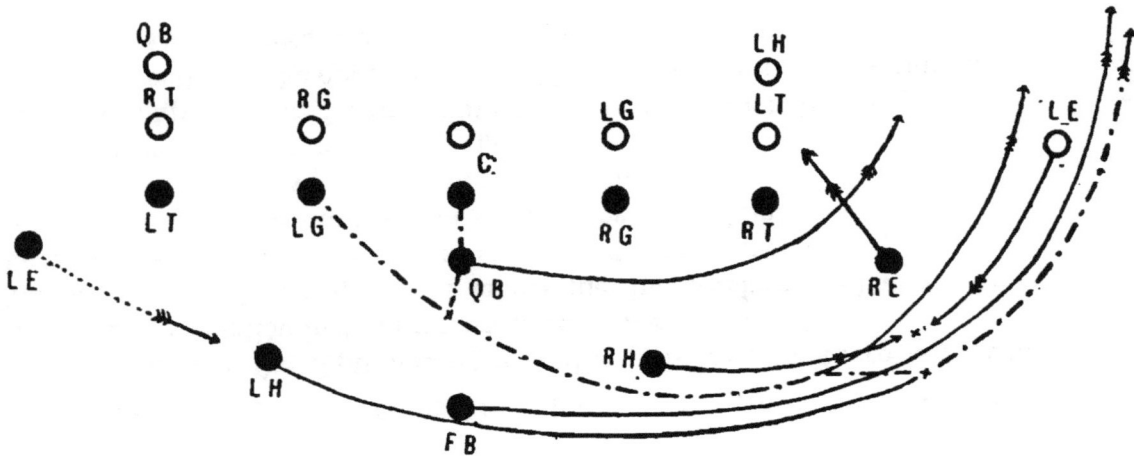

Diag. 37

37. Double pass from guard to half-back in play around the end.

To send LH around the right end on a double pass from LG, there is no change in the arrangement.

LG leaves the line the instant the ball is put in play, as shown in diagram twenty-one, receives the ball at x, and takes a direction similar to that shown in diagram twenty-three.

RE, RT, RG, and LT play as shown in diagram eight.

RH, FB, and LH start for the right the instant the ball is snapped.

RH runs directly for the opposing LE, and disposes of him. FB assists RH, if necessary, and then proceeds on down the field to interfere.

LH runs somewhat back from the line taken by FB, and as he nears the end turns back and receives the ball, on a pass of four or five yards, from LG.

LH then puts on utmost speed as he swings out around the end, while LG continues on, cutting in slightly to interfere.

LE follows directly behind LH, to make the pass safe, and to protect him from behind.

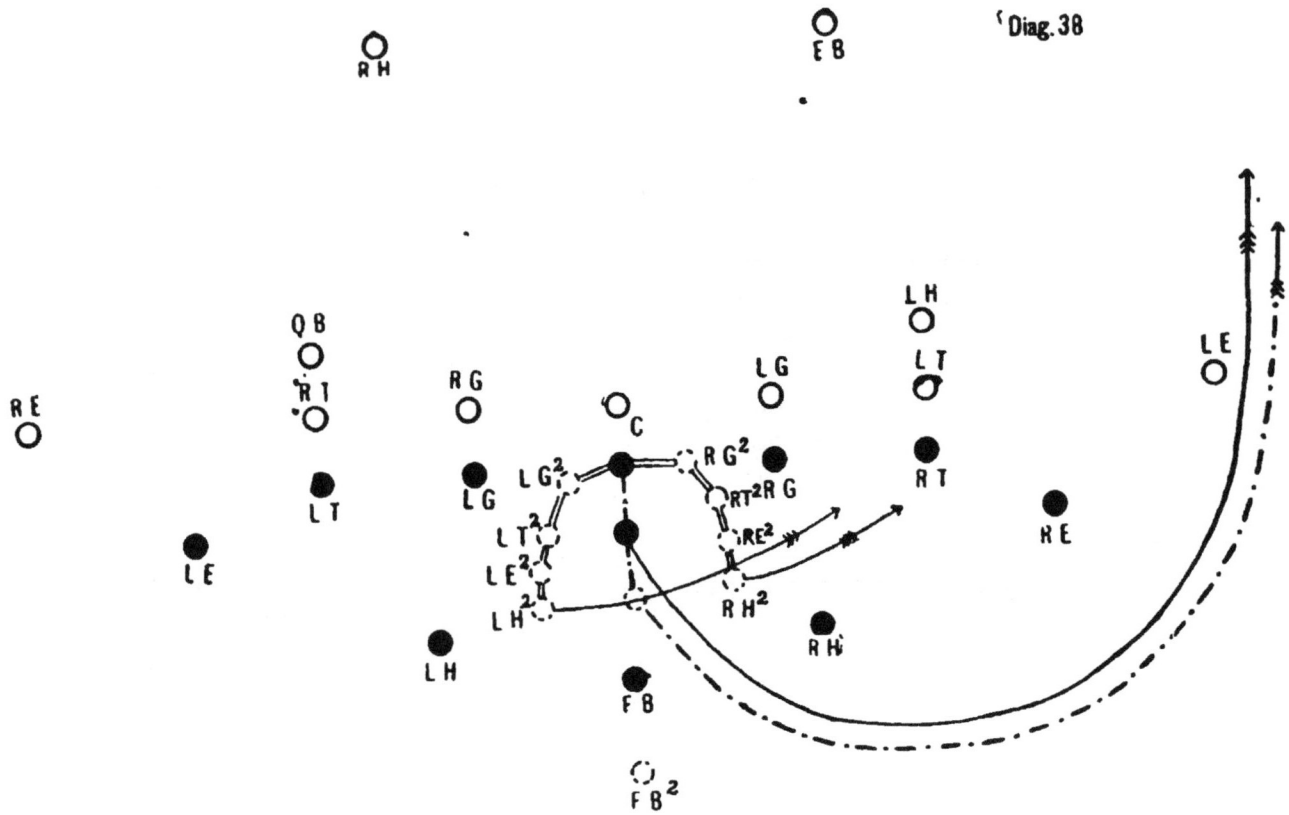
Diag. 38

38. Slow mass wedge from a down.

To send the slow pushing wedge through the center from a down, the men *spring* to their positions in the wedge formation, as shown in the cut, *the instant the signal is given*.

RG forces himself as close as possible to C's right, directly abreast of him, while LG holds himself firmly against C on the left and slightly back from the line.

The remaining rushers and half-backs take their positions behind the guards, as indicated, in a similar manner to that shown in diagram forty-one.

The men must be drilled until they can spring into their positions in the formation *instantly*. The ball should come back at the same moment and be passed to FB, who has come in to FB¹, and the whole wedge surge forward with the greatest possible force, as in diagram forty-one.

This play may be repeated several times for short gains until the tackles and ends on the opposing side are drawn well in to mass against it, when FB, accompanied by QB, will dart suddenly out from the rear for a long run around the end of the opposing team, as shown in the diagram; in which case RH and RE will cut across in the lines indicated to block the foremost men among the opponents. Should it be found that the opposing backs come up to help block the play FB may drop suddenly back to FB² and punt well down the field.

NOTE. On the play in which the FB is sent out from behind the wedge for a run around the end, there should be a little delay in snapping the ball, in order to give the opposing team more time to draw well in behind the center.

Diag. 39

39. Feint run around the end from the wedge in the line.

The wedge in the line is formed at the given signal as shown in diagram thirty-eight. The men are closely drawn into a compact formation, so that the opposing side cannot see what goes on within the wedge. The moment the ball is snapped QB slips it under the left arm of RE or RT, who receives it quietly, without making the slightest demonstration that the ball has been passed to him, and stands still in his position, bent over somewhat in the act of pushing, while the rest of the wedge plunges forward. As soon as QB has passed the ball to RE he instantly starts back from out the wedge, seizing FB by the arm as he goes. RH leaves his position at the same moment, and all three dash off to the left together, as shown in cut A, swinging in a long circle back from the line to attract attention and to give the opposing team more time to cut across the field in order to intercept them. As they pass into view LH and LE leave the wedge in the lines indicated, as if to block for them.

The opposing side at once supposes that the wedge has been simply a blind to permit the run around the end, and the entire team dash off to intercept FB, whom they suppose to have the ball. When FB and his interferers have arrived at about x in the lines indicated, RE darts out to the right unobserved.

NOTE. This play can be worked most successfully after sending the wedge straight ahead for several downs.

155

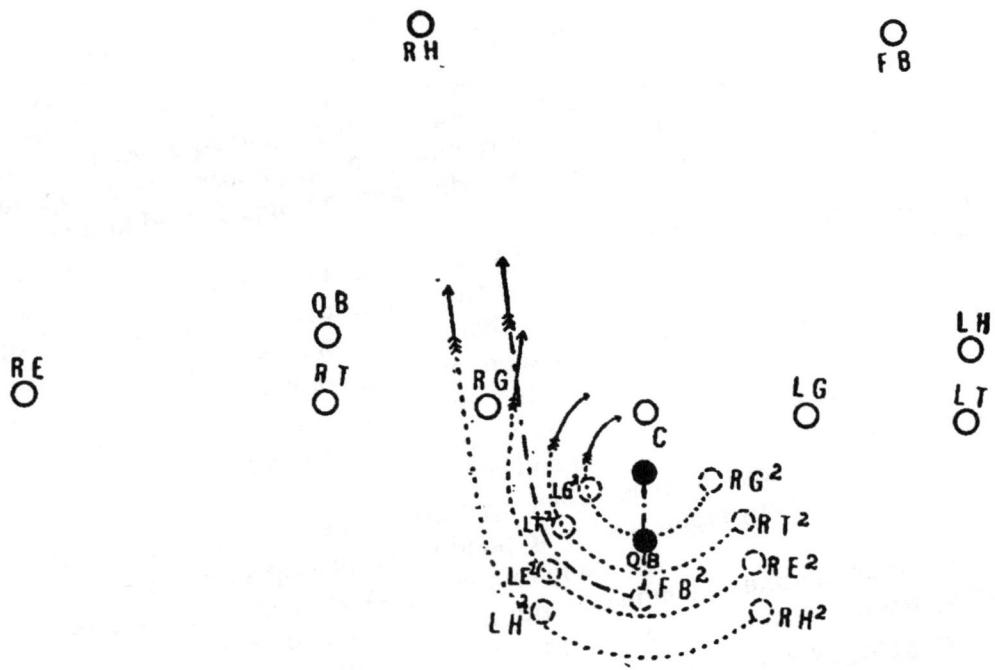

Diag. 40

40. Revolving wedge from a down.

To send the revolving wedge through the line the arrangement shown in the cut is formed in precisely the same manner as explained in diagram thirty-eight.

The ball is put in play immediately, and the entire wedge plunges *straight forward*, as before, in a *closely compact body*.

After a few seconds, when the opposing side have massed themselves in front of the wedge so that its forward progress is nearly blocked, the entire formation throws its weight to one side, each man turning slightly in order to face the direction in which he wishes to proceed, and attempts to revolve around the opposing team, *turning upon* c *as a pivot*.

The very fact that the opponents are pushing with utmost force in a direction exactly contrary to the *original* line of advance of the wedge, is of great assistance in performing the evolution.

When the wedge has swung sufficiently around, the rear men may break away and dash down the field with the ball.

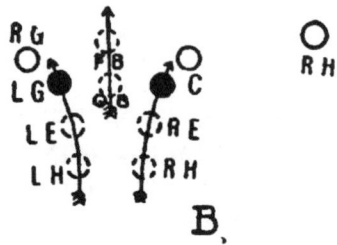

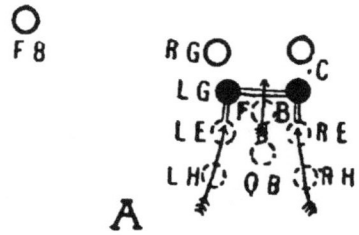

Diag. 41

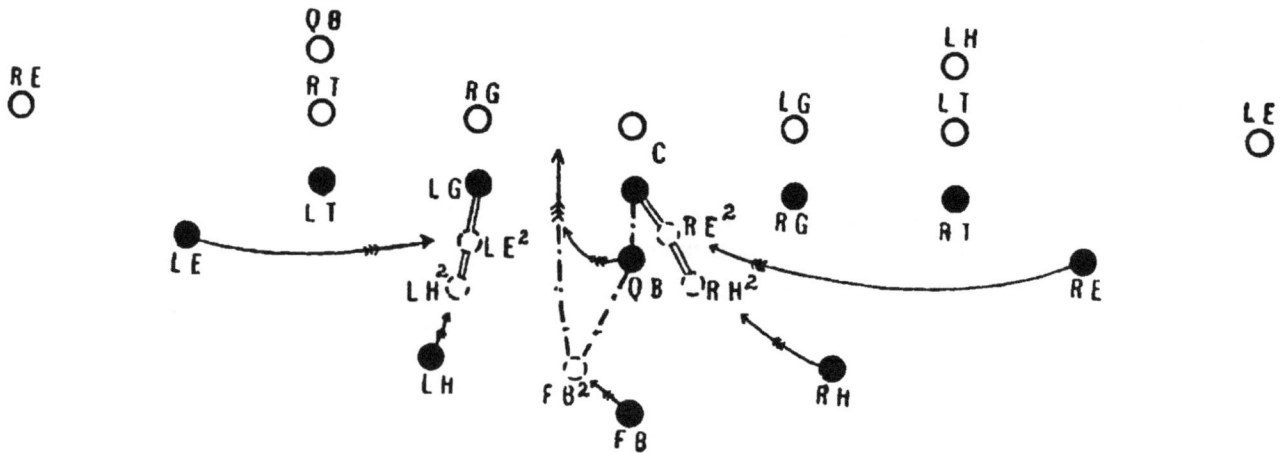

41. Lifting wedge through the center.

To send FB in a wedge through the line between IG and C, the play is peculiar in that it consists of two distinct parts. At a given signal to form the wedge, together with an additional signal which shall indicate whether the play is to go through the center or around the tackle, LE, RE, LH, and RH rush in and take the position shown in the large cut. LE stands behind LG whom he *grasps by the hips*, his arms only *slightly bent* at the elbows and his body held *well back* from LG, in the best position for pushing. LH occupies the same relative position *behind* and slightly to the *left* of LE. RE jumps in and takes the position *behind* and a little to *one side* of C, giving QB just room enough in which to receive the ball, and places his hands on the hips of C with his body braced back at arm's length. RH takes a similar position behind him and a little to the right. FB comes in slightly closer, to FB².

As soon as possible after the wedge has been formed C puts the ball in play. FB dives straight into the vortex of the wedge, receiving the ball from QB as he rushes by him, and rams his head *low down* between the hips of C and LG, the ball held tightly at his stomach with both hands.

The *instant the ball is snapped* LG and C press *close together* and *do not allow themselves to be forced apart.* FB shouts lustily "now!" the instant before he strikes the line and all *lift straight ahead* for *three* or *four* seconds, FB pushing with his head. (See cut A.) Then LG and C burst apart, carrying their men with them and allow FB and QB to shoot through the opening. (See cut B.)

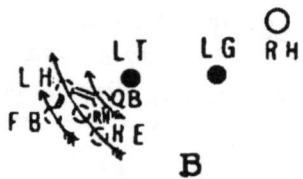

B

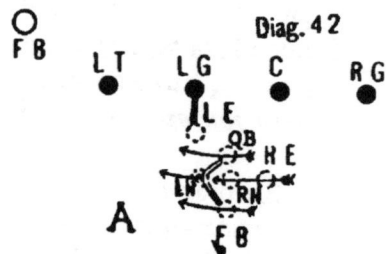

Diag. 42

A

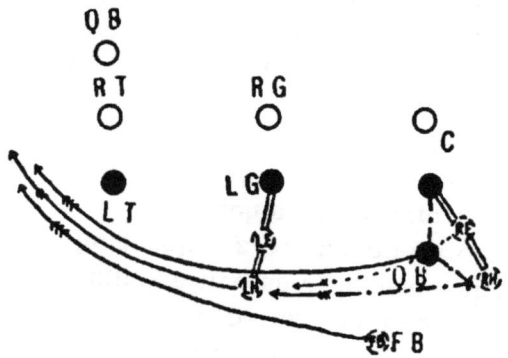

42. Wedge from the center around the tackle.

To send the wedge around the LT, a preliminary signal of "form the wedge" is given, together with the signal which is to indicate the direction of the plays, and the formation seen in the large cut is instantly made, in precisely the same manner as shown in the preceding diagram.

All the men in the wedge should have the appearance of being about to go through the center as before.

At the instant the ball comes into his hands QB whirls about, turning his *back* toward RE, and *places the ball in the hands of RH*. QB then instantly turns *back* and attaches himself to the side and a little *behind* LH, while FB, springing forward at the same moment, attaches himself in a like manner on the other side of LH, and all three dash away together in the lines indicated, around LT.* RH follows close *within* the vortex of the wedge so formed, while RE runs directly in the rear of RH and pushes with all his force as they round the tackle.† LE throws his entire weight in behind LG to help hold the line back, and LT forces his man back and to the *right*.

Note. LH must take *great care* not to leave his position until the *instant* that QB and FB reach him. The formation must be somewhat open and all endeavor to run at great speed.

Note. It must be borne in mind that the representation in the cut is diagramatical and that in reality the guards are drawn close in by the side of center, while the tackles are shoulder to shoulder with the guards in all plays with the above wedge formation.

---

* See cut A.
† See cut B.

Diag. 43

43. Quarter-back around the end from behind the wedge.

With the preliminary signal to form the wedge, a signal is given to indicate the direction of the play, and the formation as seen in the cut is taken in precisely the same manner as shown in diagram forty-one.

The instant the ball is snapped QB passes it straight back to FB, who dashes forward, hands the ball directly back to QB as he passes (being careful *not to make a forward pass*), and dives into the line between LG and C.

As QB receives the ball he instantly runs out from behind the wedge and makes for the right end at utmost speed in the line indicated.

At the moment the ball is returned to QB, RH, RE, and LH dash for the right end, LH crossing *behind* QB as he comes out from the wedge. RH runs directly for the opposing LE. RE takes the first man on the opposing side after he rounds the tackle, while LH runs on the inside of QB and interferes for him as they go together down the field.

LE follows in the rear of QB to protect him and prevent him from being caught from behind.

LT holds his man hard, and LG, RG, and RT play as shown in diagram nineteen.

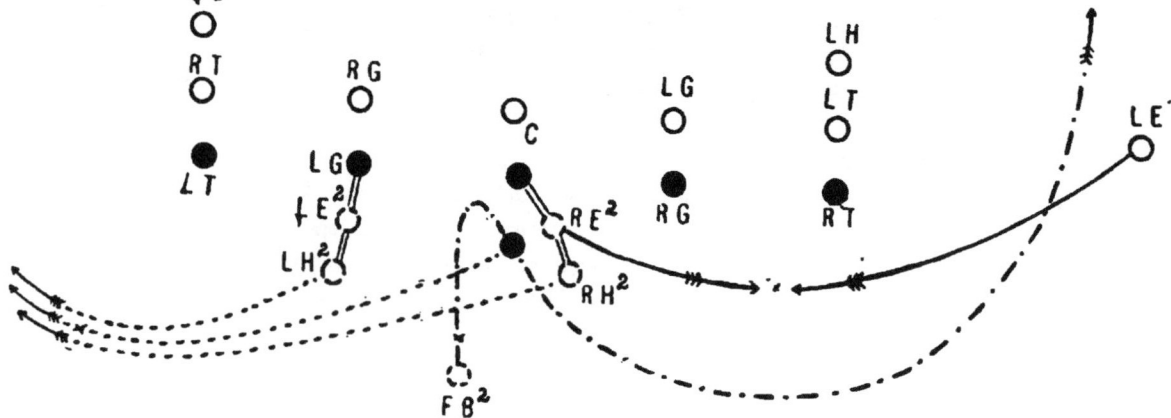

Diag. 44

44. Feint play from the wedge.

At the signal the wedge is instantly formed as shown in diagram forty-one.

As the ball is snapped FB rushes forward and, receiving it from QB at x, immediately crouches down behind c, shielded from view by the center, guards, and ends, who are *tightly massed together.*

QB then instantly darts out from behind the wedge to the left, accompanied by LH and RH, who swing well back from the line to attract attention, and hold closely together * to deceive the other side into the belief that they have the ball.

The opposing team immediately suppose that a play around the end from the wedge is being attempted and rush toward the side of the field to intercept it.

FB remains crouched behind c until QB, RH, and LH are well off toward the side of the field, as at x, and then suddenly springs up and circles the opposite end, accompanied by RE, who makes sure to block the opposing LE in case he has come around behind the line.

NOTE. Compare this play with that shown in diagram thirty-nine.

---

* See cut A.

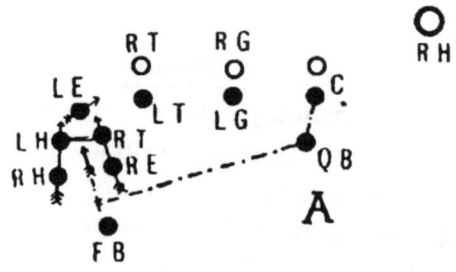

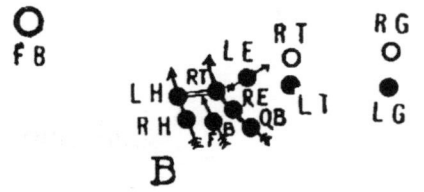

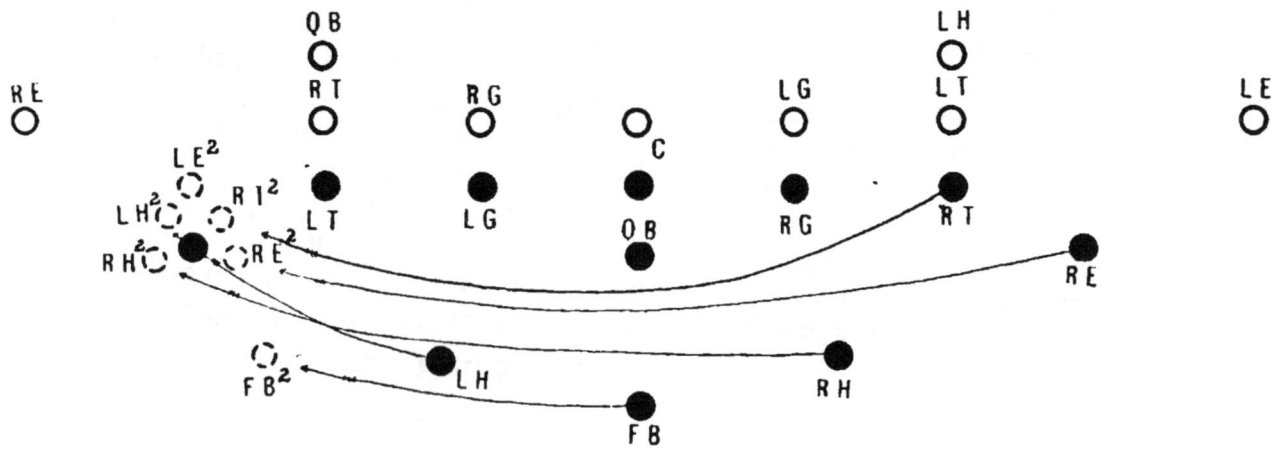

Diag. 45

45. Wedge on the end of the line.

To form a wedge on the left end, at the given signal LE comes up into the line, and RT, RE, LH, and RH rush on instantly and form the wedge directly behind him, while FB moves over to a position about two yards in the rear of RE, as shown in the large cut. C allows RT just sufficient time to reach the left end, by which time the wedge will be perfectly formed if the men are properly drilled, and then snaps the ball as soon as possible. As QB receives it he springs toward FB, passing him the ball as he does so, avoiding all possibility of being caught by the opposing LT. FB plunges forward at the same moment, receiving the ball at x (see cut A), and at the same time shouts "Now!" as he rushes in behind the wedge. At that same instant the whole wedge dashes forward *on a slight angle to the left*, LE jumps into the opposing RT, or the extra man in the line, while QB attaches himself to the rear of RE as the wedge rushes forward (see cut B). FB *must* succeed in getting well in between RH and RE, while all rush forward with utmost force, LH and RT *holding firmly together*.

NOTE.[1] RH may leave the wedge to take the opposing RE if he attempts to break in from the side.

NOTE.[2] In case the opposing side sends the backs up into the line to mass against the wedge and block it, FB may kick the ball down the field instead of rushing it, QB protecting him from the opposing LT as he does so. While not an especially strong position from which to kick, a short quick kick just over the heads of the opposing back will serve every purpose, as on all future similar formations it will compel the opposing side to retain at least one man well behind the line as a protection.

Diag.46

46. Play around the opposite end from the wedge on the end.

After the wedge has been formed on the end, as shown in the preceding diagram, there may be some delay in snapping the ball, and the opposing LT, or in case of an inexperienced team, both LT and LE run around in order to mass against the wedge when it advances.

In that case the following modification may be made. Before the ball is snapped the captain, seeing that the opposing LT has run around, gives some key word which all understand as a signal to indicate that the play is to be changed to one around the opposite end.

QB makes the pass as shown in diagram forty-five, but passes to RE instead of FB, and then instantly turns and precedes RE around the right end.

FB and RE dash toward the right as the ball is passed, FB preceding and blocking the opposing LE.

RG attempts to lift his opponent back and force him in toward the center as far as possible.

As RE starts forward in the line indicated, LG breaks away from his man as shown in diagram eight, and preceding RE, dashes into the LG on the opposing side in case he succeeds in getting around RG, while RT jumps in and takes the man whom LG has left exposed. RH follows directly in the rear of RE to prevent him from being taken from behind, while LH and LE block the extra men in the line.

NOTE. This maneuver will prevent the opposing team thereafter from drawing all their men from the opposite end to mass in front of the wedge.

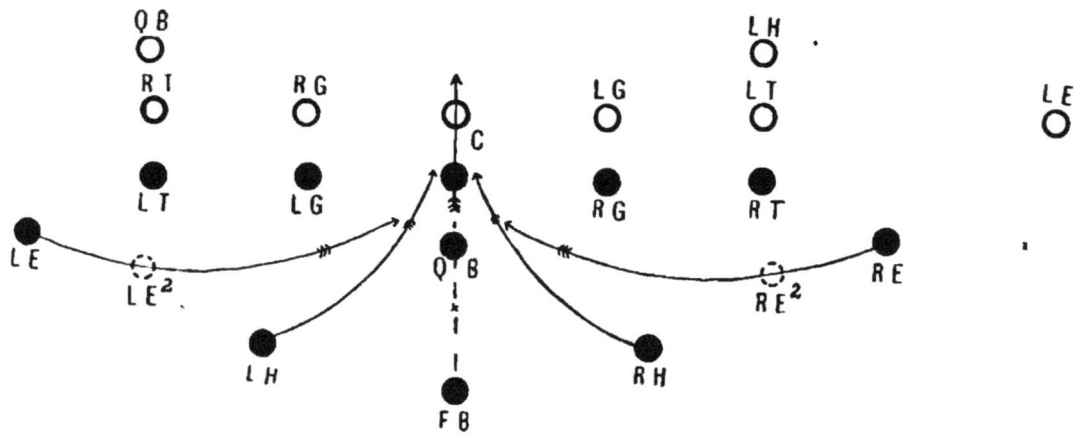
Diag. 47

47. Running mass wedge through the center.

To send FB through the center on a running mass play directly behind c, the ends and backs start forward the instant the ball is snapped.

The guards lift their men *back* and *out from the center*, while c endeavors to force his man straight ahead of him.

LH and RH dash in and attach themselves behind c on each side of him.

FB springs forward at the same moment, and receiving the ball on a pass at x from QB, who slips to one side, dives in directly behind c with *head down*. At the same instant that FB reaches the line, the ends close in on either side of him directly behind the half-backs. QB throws himself in the rear of FB, and all push forward with the greatest possible force in a solid and tightly formed mass.

The vital point in the play is that all strike the line at as nearly as possible the same instant and form a *tightly massed wedge*, which is driven directly through the line.

NOTE. By drawing the ends in to LE^2 and RE^2, they may be enabled to strike the line ahead of the half-backs, in which case the latter will attach themselves on either side of FB as he rushes forward. The wedge must never cease pushing until the man with the ball is actually downed and absolutely held.

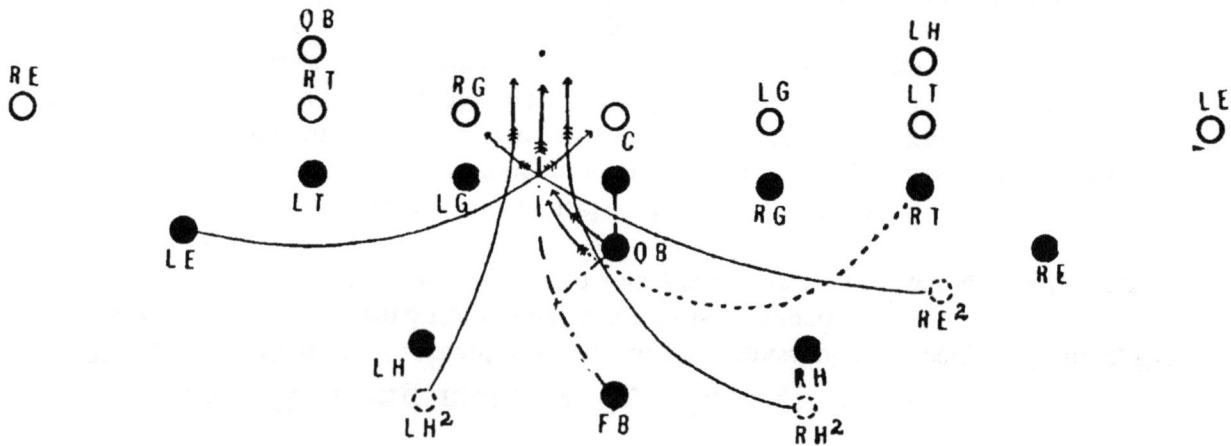

Diag. 48

48. Running mass wedge between guard and center.

To send the running mass wedge through the line between LG and C, the halfbacks draw back slightly before the ball is snapped to LH² and RH², in a line with FB, in order to give the ends more time to reach the opening ahead of them, and also to enable themselves to gain greater headway before striking the line.

RE also works over slightly to the left to RE².

At the instant the ball is snapped, all the men behind the line dash straight for the opening in the lines indicated.

C lifts his man back and to the *right*, and LG forces his man back and to the *left*.

LE passes through the opening ahead,* at an angle, and strikes the opposing C with his full force, while RE, crossing directly behind him strikes the opposing guard in a similar manner.

At the same moment, FB with his head down and the ball held as before, strikes the opening so made, immediately behind the ends, with the greatest possible force, the half-backs firmly attaching themselves to his flanks, as he receives the ball at x, and forcing him through. RT leaves the line as the ball is snapped, as shown for LT in diagram thirteen, and together with QB, closes in behind FB. All mass firmly together as before and drive directly through the line.

LT and RG hold their men and force them *out*.

---

* When the ends find difficulty in reaching the opening ahead, they may follow the half-backs as in the preceding play.

Diag. 49

49. Running mass wedge directly at the guard.

To send the running mass wedge into the line directly behind LG, very little change is made from that shown in diagram forty-seven except that the play is directed at the guard instead of C.

C and LT force their men back to the right and left respectively, as the ball is snapped; the half-backs dash forward in the lines indicated, and attach themselves to the flanks of LG with the greatest possible force; the ends strike the formation so made, at that same instant, and each grasps FB from the side as he rushes in behind LG with his head down; while RT and QB throw their weight in behind FB as shown in diagram thirteen, and all plunge forward in a mass pressed firmly together.

The success of the play depends upon the formation of a compact mass which continues to hold firmly together after it strikes the line, and in which all push with their whole weight.

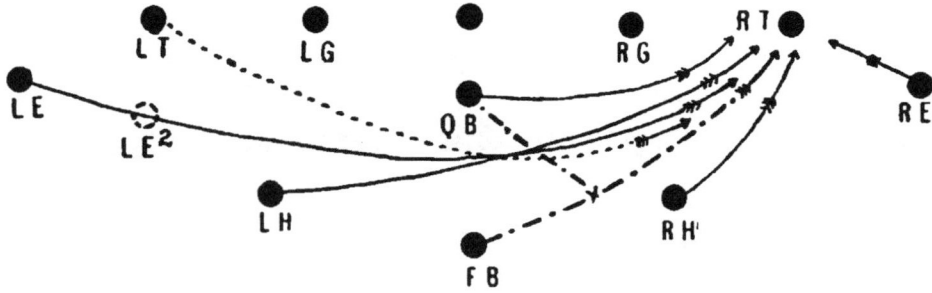

Diag.50

50. Running mass wedge directly at the tackle.

To send the running mass wedge into the line behind the tackle, the same principle is carried out as that shown in diagram forty-nine.

The instant the ball is snapped all the men behind the line start for RT in the directions indicated, at utmost speed, FB receiving the ball at x on a pass from QB. LH and RH strike the flanks of RT at full speed on either side, while the ends, QB and LT, mass on the sides and rear of FB as he strikes the line, as shown in diagram forty-seven.

The mass must continue its closely locked formation and push directly down the field.

LE works in slightly to LE^2 before the ball is put in play.

LT leaves the line the instant the ball is snapped, as if he himself were to run with the ball, follows directly behind FB and overtakes him if possible before he reaches the line, in order to push him as he strikes.

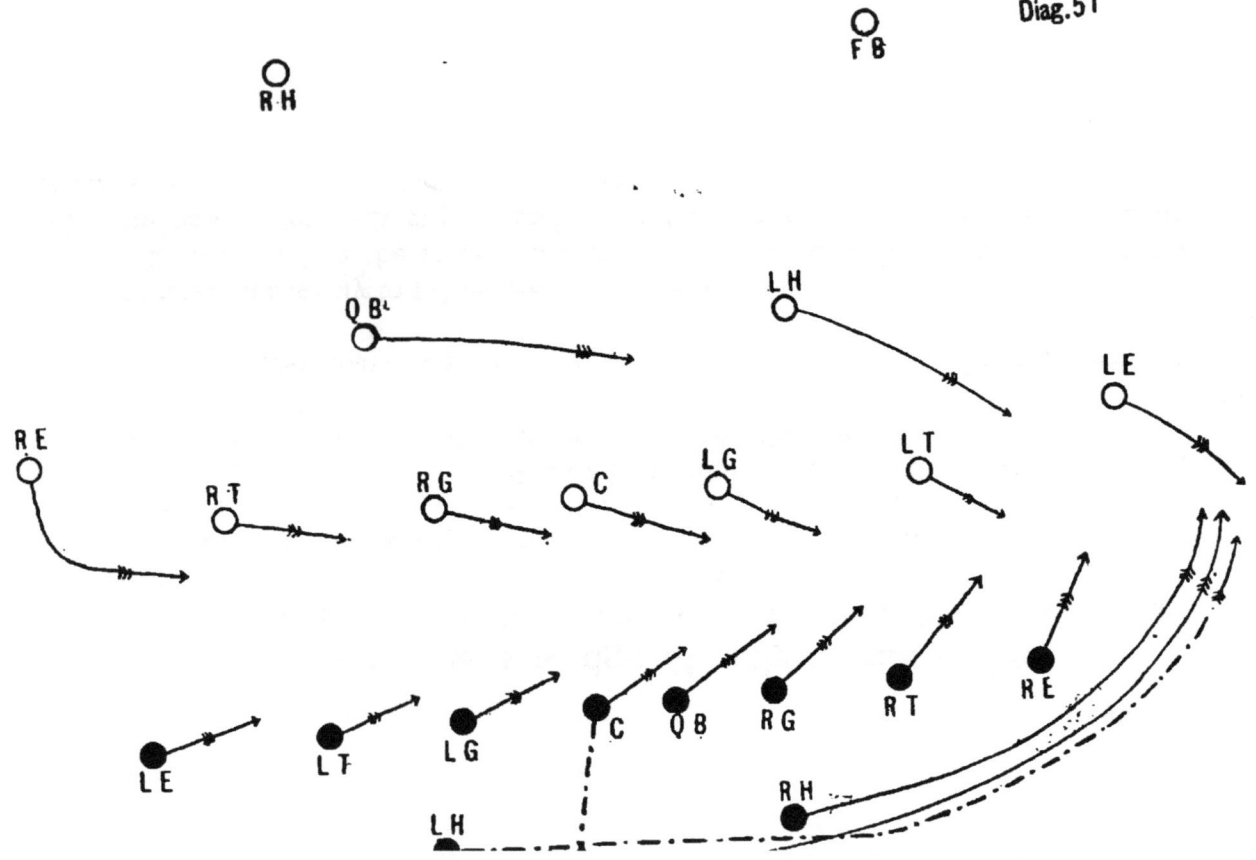

Diag. 51

51. Free opening play from the center of the field.

The men are lined up at the center of the field as shown in the cut, from five to six feet apart.

A signal makes it understood around which end the play is to be made, and each player selects the man on the opposing side whom it has been prearranged that he shall block.

C puts the ball in play by kicking it while still retaining it in his hands, and passes it back to LH, in case the play is to be made around the right end. At the same instant the entire rush lines move diagonally toward the right as one man, and dash into the opposing rushers as they meet midway between the two lines.

RH and FB precede LH to interfere for him as they reach the end, and all sprint at topmost speed.

NOTE. The rushers must see to it that they do not betray by their looks, before the ball is put in play, the direction in which the run is to be made.

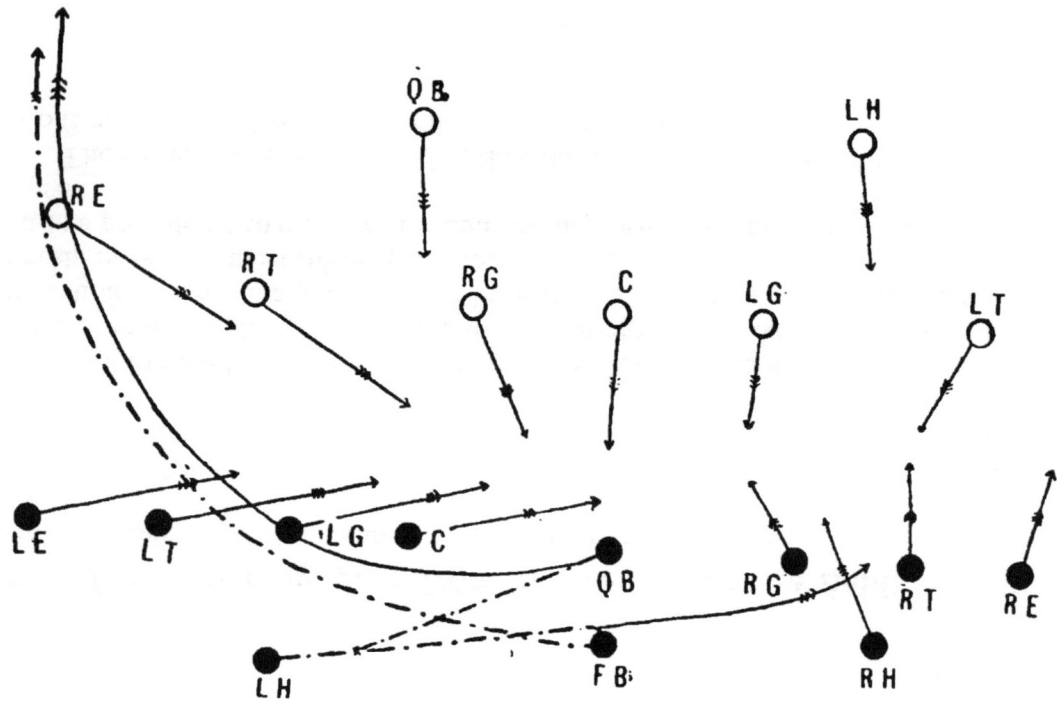

52. Double pass opening play from the center.

To make the double pass opening play around the left end, the rushers are placed as indicated, on the center line, about two yards apart, leaving an interval of about ten yards between c and QB, and five yards between QB and RG, while the backs are about three yards behind the line.

The instant that QB puts the ball in play* c, LG, LT, LE, and LH dash toward the center of the field in lines nearly parallel with the cross lines, preserving their distances from one another. QB makes a long pass to LH at x, as he advances toward him on the run.

The men to the right of QB start directly down the field, swinging in slightly toward the center to block the oncoming rushers. QB and FB stand still.

LH passes close in front of FB, carries the ball in his right arm, and passes it to FB, to take it as he rushes by.

FB and QB then instantly start in the opposite direction and sprint at utmost speed to encircle the opposing team which has been drawn in toward the center.

NOTE. From this same formation QB may pass the ball either to FB or RH for a kick, in which case the rushers will all run straight *down* the field instead of across in the lines indicated in the cut.

---

*See description, diagram fifty-four.

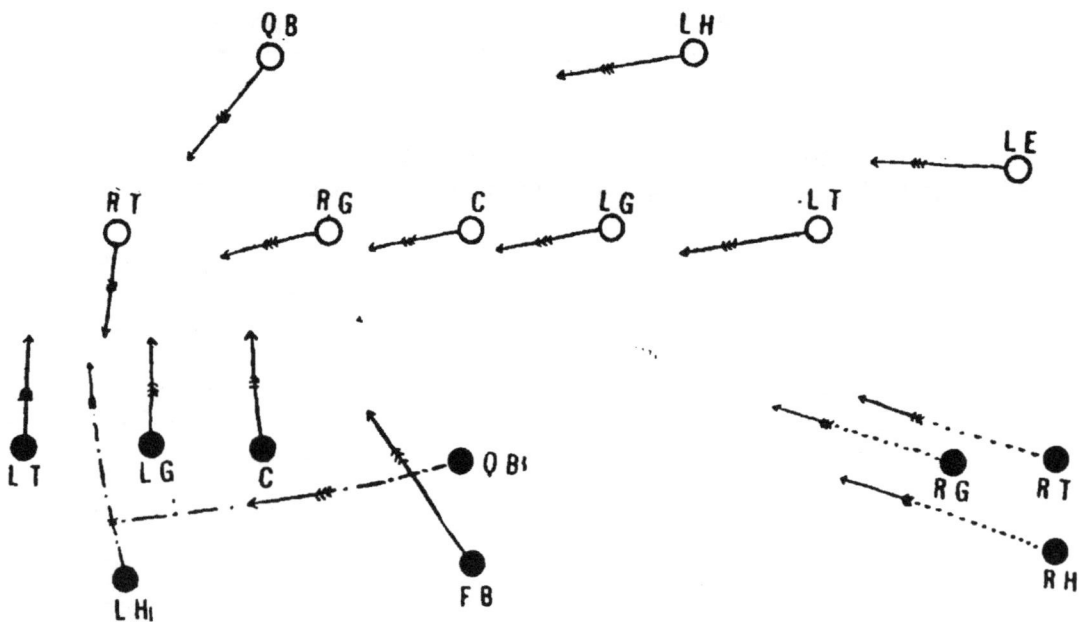

Diag. 53

53. Opening play from the center with team divided.

The men are arranged as indicated in the diagram, on either side of the field, the rushers being about two yards apart and the ends about five yards from the side lines. F ʙ is placed about two yards, and the half-backs about three yards, behind the center line.

Q ʙ looks to the one side and then to the other, in order to render the opposing team uncertain whether the ball is to be passed to the right or left, or to Fʙ for a kick down the field. He then puts the ball in play as shown in diagram fifty-one, and makes a long pass to ʟ ʜ, who receives it on the run at x, following it immediately to make the play safe, in case of a wild pass.

The instant the ball is in play every player in the field will dash forward in the lines indicated, *except RE*, who stands still in his position or drops flat upon the ground close to the side lines, unobserved by his opponents as the other three men dash across the field.

It is of no consequence if only a small advance is made. All depends upon the quickness of the following play for success. Every man upon the team rushes to his position in the line, and without waiting for a signal the ball is immediately snapped and a long pass made by Qʙ straight across the field to Rᴇ, who catches the ball upon the run and has the entire field before him.

Nᴏᴛᴇ. This play is only practical when Qʙ can be relied upon for a long and accurate pass, and when the wind is not unfavorable.

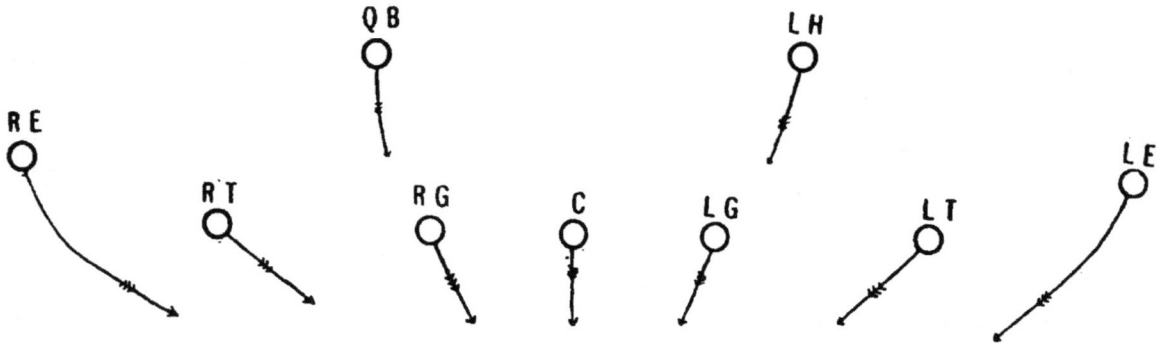

Diag. 54

54. *Princeton opening wedge from the center of the field.

To send the wedge straight down the field from the center, the men form in the positions shown in the cut, as closely and firmly bound together as possible.

At the signal, c puts the ball in play by touching it with his foot, and passes it back to QB, who is immediately behind him, ready to receive it. As the ball is put in play the entire wedge rushes forward in a compact mass, preserving its formation, and endeavors by mere force of weight and momentum to advance the ball as far as possible straight towards the opponent's goal.

QB upon receiving the ball places it at his stomach, clasps it tightly with both hands and charges forward with head down, while FB throws in his entire weight from behind.

NOTE. It is a vital point that all the men keep their feet and run in a compact mass, preserving the formation.

---

* The wedge formation at the center of the field originated at Princeton.

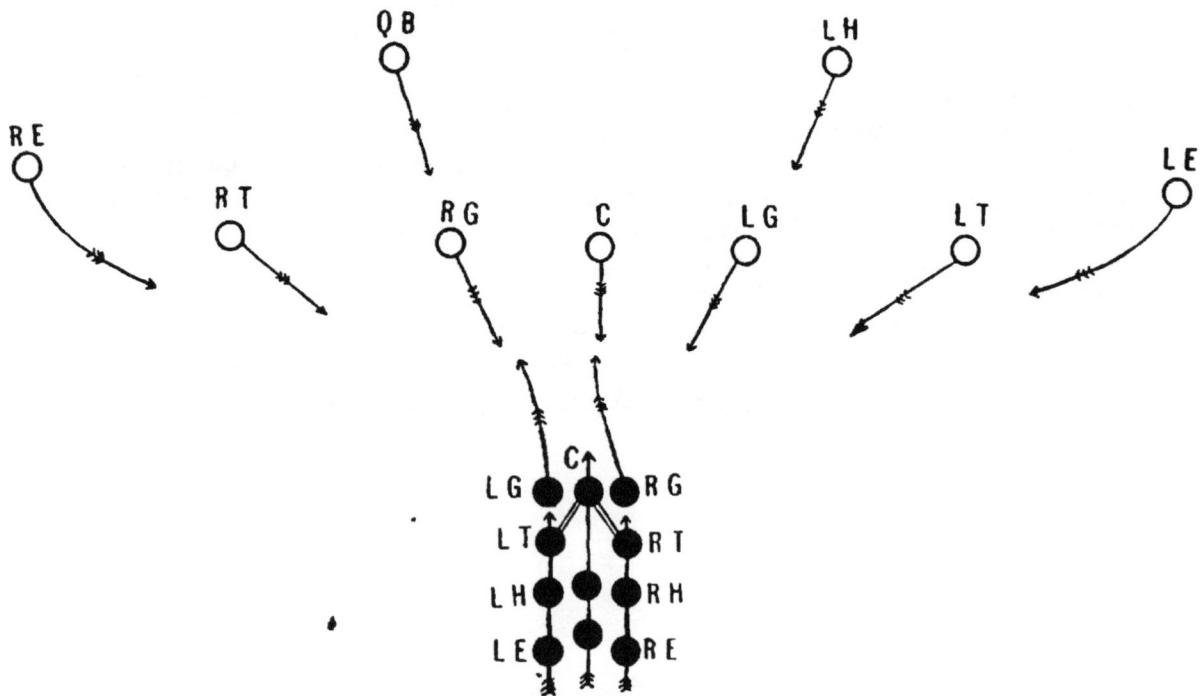

Diag.55

55. Yale modification of Princeton wedge.

This wedge differs in one very important respect from the preceding. Instead of being a part of the wedge formation the guards are placed *outside* of the wedge directly abreast of c.

The instant the ball is put in play, LG and RG spring forward in advance of the wedge, and meet the opposing guard and center *midway between* the wedge and the point from which their opponents start.

LG jumps directly into his man and attempts to throw him to the *left*, while RG, meeting the opposite c in the same manner, attempts to throw him to the *right*.

The wedge advancing immediately behind is thus saved the shock of being struck by the opposing guards under full headway.

The wedge may charge thus at an angle slightly to the right or left, the guards taking the opposing c and RG or c and LG, as the case may be.

NOTE. As it is highly desirable that the men without the wedge be swift and dashing, it may be found more advantageous to place the tackles, or two comparatively light men, in these positions, while the guards are retained within the wedge itself.

Diag.56

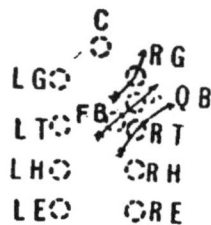

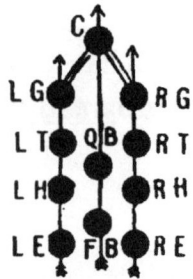

56. Princeton split wedge.

The formation is precisely the same as that shown in diagram fifty-four. The ball is put in play as before, and the wedge advances straight down the field.

As the on-coming rushers strike it, the wedge suddenly opens at some point previously agreed upon, and allows QB, who carries the ball, to break through and dart down the field.

The opening usually selected is that between guard and tackle, as shown in cut A ; in this case, the guard and tackle separate and force their opponents to the left and right, respectively, while QB, with his head down, and FB pushing him from behind, forces his way through, and breaks clear of the wedge.

This opening may be made either to the right or left, and at any point desired.

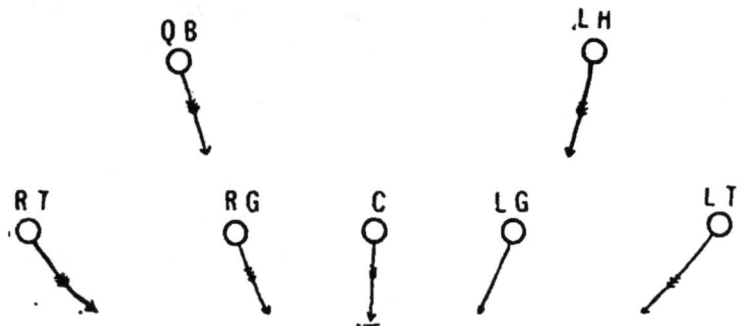

Diag. 57

57. Yale split wedge.

The formation is very similar to that of diagram fifty-six, but the two lines are arranged so that the parallel sides are *nearer together*. The *half-backs* and *ends* brace themselves well back, at arm's length from the men directly ahead of them, precisely as shown in diagram forty-one.

QB stands well back between the half-backs, as shown in the cut. The guards, tackles, and half-backs stand with toes pointing *straight forward*, to leave a narrow unobstructed lane between the two lines, down which the ball is to be rolled *on its side*.

C places the ball on its side between his legs, and puts it in play by touching it with his toe (the ball the while firmly held under his hand), and rolling it back. At the same moment the entire wedge surges straight ahead. QB will have had just time to secure the ball, turn toward RH, when the charging rushers will be upon the apex of the wedge. LH then instantly turns *square to the right*, and seizing RH by the arm, knocks him directly out of the wedge on the opposite side. LE follows immediately behind LH, FB attaches himself to RE, and the *new* wedge, with RH as its apex, and QB directly in the center with the ball, dashes off at an angle of about 45° to the original direction. (See cut A.)

NOTE. The wedge may split either to the right or left. It is very important that the second wedge preserve a *loose*, yet firm, formation, so that all may run at utmost speed.

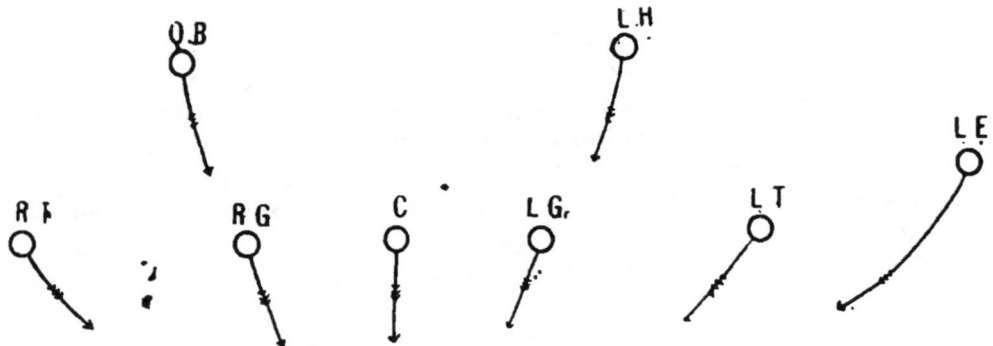

Diag. 58

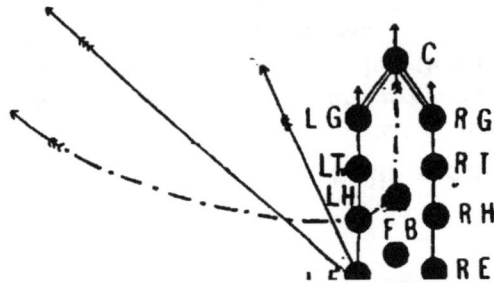

58. Side play from wedge at the center.

The formation is precisely as shown in diagram fifty-four. C puts the ball in play as before, and passes it back to QB, and the entire wedge advances with a rush.

QB quietly transfers the ball to LH unperceived, under cover of the wedge formation; and as the on-coming rushers strike the apex, LE and FB, who has changed his position to FB², slip by on the side in the lines indicated, to block the opposing RT and RE, while LH darts suddenly to the left, passing *behind* LE and FB, and attempts to encircle the end, sprinting at utmost speed.

NOTE. The ball may be passed by QB to either of the half-backs or ends, and the play made as shown.

A play is sometimes made closely allied to this, in which QB remains in the wedge until after it has encountered the opposing rushers, and then suddenly darts out from behind, with a single interferer, trusting to his speed and the unexpectedness of his appearance to carry him safely around the opposing team.

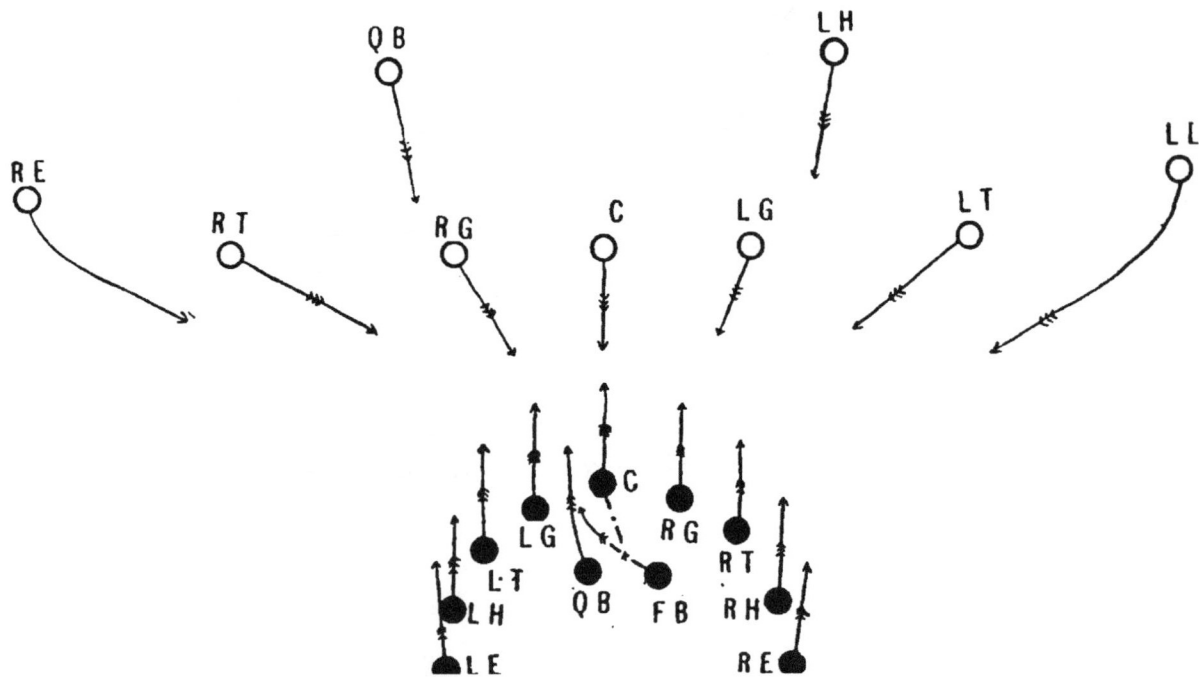

Diag. 59

59. Hard running wedge with loose formation.

The men are formed in a wedge shaped as above, from three to four feet apart, with the half-backs in the center.

C puts the ball in play by kicking it and passing it back to one of the half-backs, and the whole formation dashes forward at utmost speed without massing together, preserving the arrangement as far as possible.

Each man in the line dashes directly into the opposing rushers as they meet the wedge, while the half-back with the ball, assisted by his fellow, endeavors to slip through the most favorable opening which the open formation shall offer.

The play may be directed straight ahead, or to the right or left.

NOTE. If the entire formation, without changing its arrangement, withdraws ten yards behind c, and comes dashing forward on the run, c putting the ball in play just before it reaches him, the play may be made even more effective.

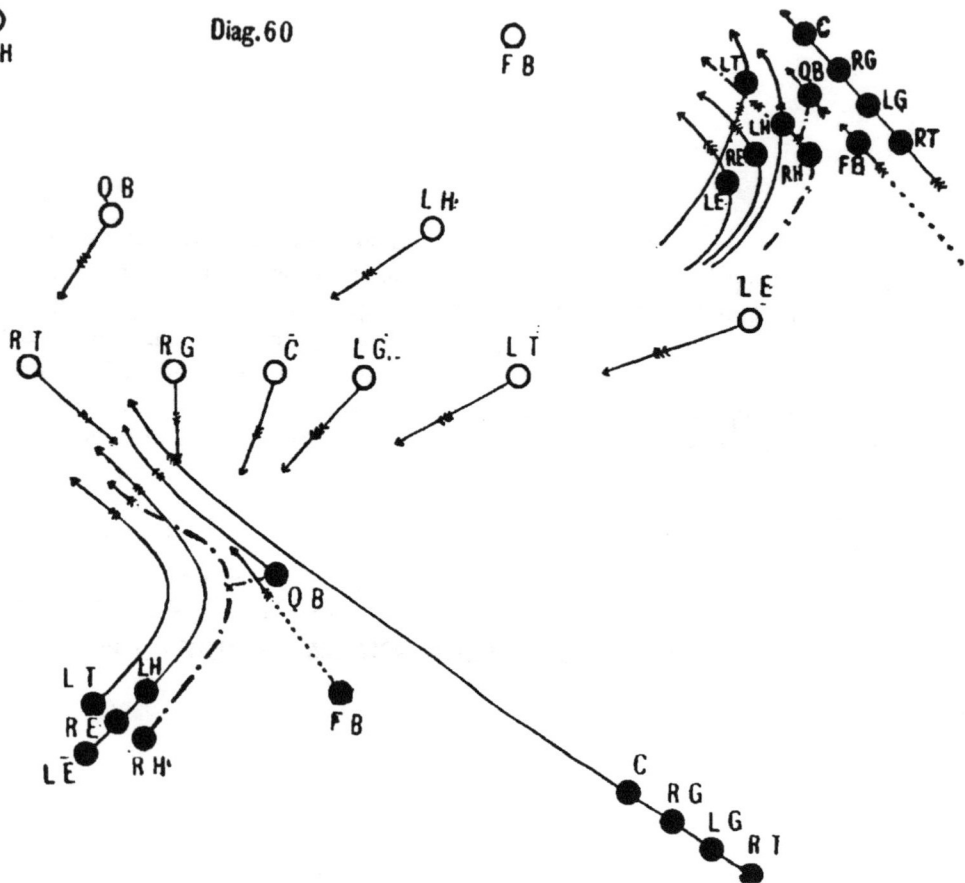

Diag. 60

60. The Harvard flying wedge.

QB stands with the ball in the center of the field. FB stands from five to ten yards behind QB and a little to the right. The remainder of the team is divided in two sections.

Section No. 1 is composed of the heaviest men in the line and is drawn up from twenty to thirty yards from the center, back and to the right, facing QB.

Section No. 2 is composed of the lighter and swifter men, drawn up five or ten yards back and to the left of QB.

Section No. 1 has the "right of way," the others regulating their play to its speed.

At a signal from QB, section No. 1 dashes forward at *utmost speed* passing close in front of QB.

At the same moment FB and section No. 2 advance, timing their speed to No. 1.

Just before the sections reach the line QB puts the ball in play, and as they come together in a flying wedge and aim at the opposing RT, or straight down the field, passes to RH and dashes forward with the wedge.

A slight opening is left in front of QB to draw in the opposing RT. (See small cut.)

As opposing RT dives into the wedge, LH and QB take him. RE and LE swing out to the left to block opposing RE. At the same moment RH puts on utmost speed and darts through opening between LH and RE.

NOTE. The arrangement of the men is arbitrary. The wedge may be directed against any point desired. Its strength lies in the fact that the men are under full headway before the ball is put in play.

Diag.61

61. Guard drops back and bucks the center.

To send RG to buck the center, at a given signal RG runs back from the line and takes the position at RG^2, while RH jumps in and fills temporarily the position of RG.

As soon as RG is in his position the ball is snapped. RG dashes forward, receives it at X, and plunges into the opening to the right of C with his head down, striking the line *hard*.* An opening in the line is made as shown in diagram five.

RE, FB, LH, LE, and QB all rush in behind RG, starting forward in the lines indicated as the ball is put in play and push, as in diagram twenty-one.*

LT, LG, and RT play as in diagram five.

NOTE. The above play is valuable for a light team if they happen to have a heavy and powerful guard.

After RG has been sent at the center once or twice it will be very effective to have QB pass the ball to LH, instead of RG, for a run around the right end, as shown in diagram eight. RG will then precede LH as does RH in diagram eight.

---

* See NOTE, diagram one.

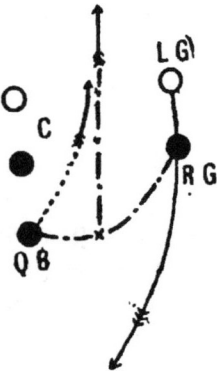

62. Guard through his own opening on the same side.

To send RG through his own opening to the right of C, the full-back and half-backs are placed back as if for a kick, as shown in the cut. FB assumes every appearance of being about to receive and kick the ball.

As the ball is snapped RG jumps slightly back and toward QB, allowing the opposing LG, who is eager to break through and stop the kick, to pass through the line *outside* of him. QB instantly hands the ball to RG, who plunges back into the line through the opening left vacant by the opposing guard, with QB directly behind him.

C endeavors to force his man well to the left as he snaps the ball.

A similar play is less successfully attempted at the tackle.

The old play of having the end lie well out and receive the ball on a long pass from QB is now almost absolutely discarded.

NOTE. Care must be taken by QB not to make a forward pass.

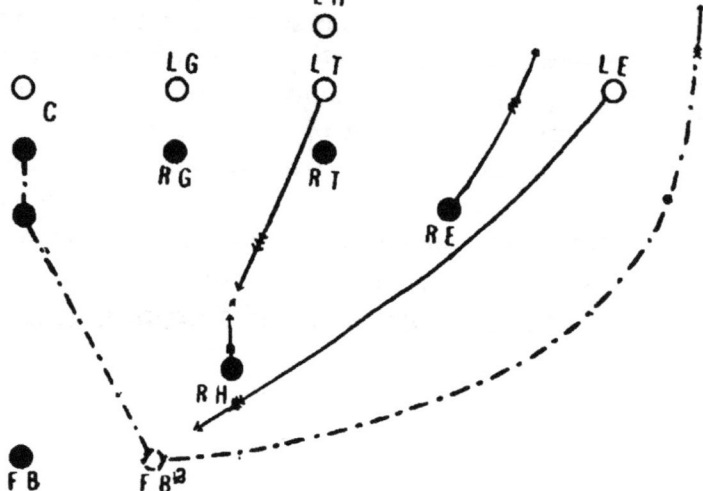

Diag. 63

63. Full-back feints a kick and runs around the end.

To send FB around the end on a feint to kick, RH and LH draw from one to two yards behind their original positions, while FB moves over toward the right to FB², from two to three yards in the rear of RH.

When the ball is put in play QB passes carefully and accurately to FB, who, with coolness and *deliberation*, without betraying by the slightest glance or uneasy movement that he is about to run, goes through the preparatory movements of being about to kick.

All the men in the line block their opponents as usual, with the exception of RE. The latter allows his man unimpeded progress straight for FB, simply forcing him to run to the *inside* of him as he passes, or takes the first extra man in the line outside of RT.

As the opposing LE is *almost upon him*, coming forward at full speed, FB suddenly darts to the right in order to dodge LE, which is easily done, and dashes around the right end. The entire success of the play will depend upon the coolness and skill of FB in waiting until the last moment before dodging to the right and in not allowing his ultimate design to be prematurely discovered. In case the opposing tackle succeeds in breaking through the line RH must take him and force him to the *inside*.

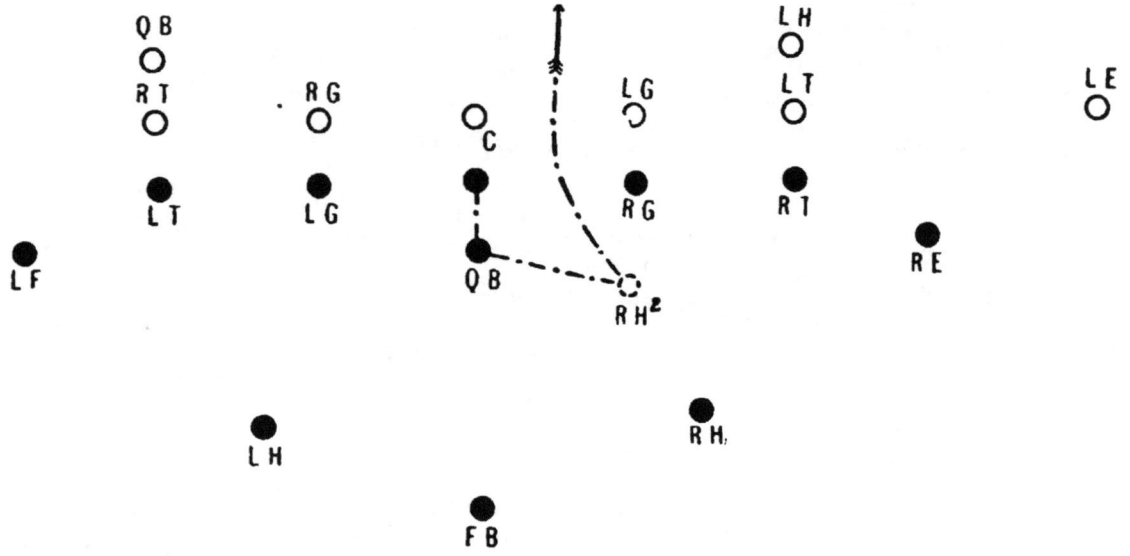

64. Full-back feints a kick and half-back darts through the line.

To send RH through the line on a feint to kick, LH and FB drop back. RH remains in nearly his original position, while FB assumes every appearance of being about to receive and kick the ball. Just before the ball is snapped RH draws in slightly nearer to RH². Upon receiving the ball QB instantly passes it to RH, who is close to the line and plunges directly through the opening between C and RG with head down, and the ball held as shown in diagram five.

The opposing guards and center are all intent upon breaking through and stopping the kick, and are entirely unprepared for a dash into the line.

As the ball is put in play RG throws his man suddenly and violently to the *right*, while C throws his man in a similar manner to the *left*, and RH darts through the opening so made without assistance or interference.

NOTE. This play is also frequently made between guard and tackle.

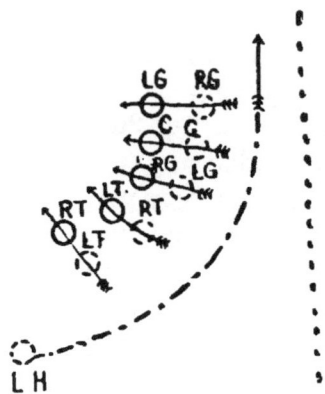

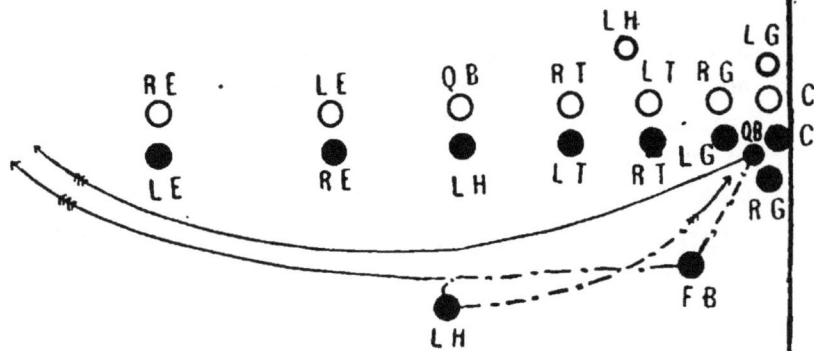

Diag.65

65. Criss-cross play from the side line.

To perform the criss-cross play when the ball is out of bounds at the side line, c places one foot well within the field, keeping the other out of bounds, and faces the opponents' goal with the ball in his hands ready to put it in play by touching it to the ground and passing it back between his legs to QB. RG stands as near the side line as possible directly behind c, out of the way of QB.

FB stands about two yards behind QB, while LH occupies a position three or four yards behind the line and from ten to fifteen yards from the side of the field. The remainder of the men stand closely and solidly together in the line.

When the ball is put in play QB passes quickly to FB and both start at utmost speed for the center of the field in the lines indicated. FB runs close in front of LH and gives him the ball as he passes, upon which LH instantly starts back in the opposite direction.

Just before FB reaches LH, the guards and center swing around to the side and sweep their opponents a yard or two into the field, leaving a narrow lane by the side line down which LH may pass, as shown in cut A.

The play can be made with equal success when the ball is " down " within a yard or two of the side lines.

NOTE. LH must take the greatest care not to step over the boundary line as he runs.

14

Diag.66

66. Harvard line wedge.

To form the line wedge upon the right side. At the given signal RH, RE, LH, and LE, dash toward the right and form a wedge directly behind RT, occupying the positions RH², RE², LH², and LE², and taking a formation similar to that shown for LE, LH, RE, and RH in diagram forty-one.

C puts the ball in play the instant the men reach their positions, and QB passes to FB, who receives the ball at x, on the run. LT leaves his position the instant the ball is snapped, and follows directly behind FB. Upon receiving the ball, FB dashes in behind the wedge with head down, and all plunge forward, preserving a compact mass. LG blocks his man hard, while C and LG block their men and endeavor to force them to the left.

As soon as QB has passed the ball, he should dash forward to throw his entire weight behind the wedge.

See NOTE, diagram sixty-eight.

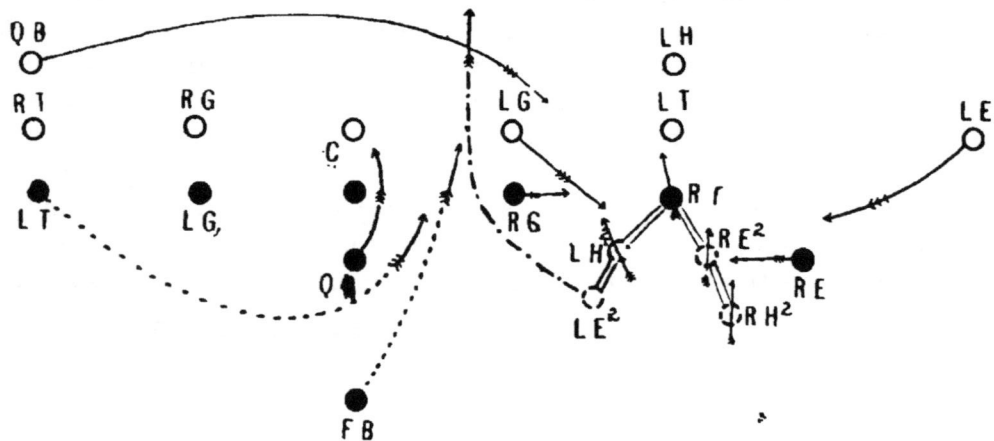

Diag. 67

67. Left end between guard and center from the line wedge.

This play is a modification of the one shown in diagram sixty-six.

When the wedge has been sent forward several times in succession for short gains of from two to five yards, and the opposing LG has found the way to dive into it low down between RT and LH, to stop it; at the signal for the play, RG allows his opponent to break through to the right, as the ball is snapped, without resistance, and then forces him further to the right. C blocks his man and forces him hard to the left.

QB then instantly passes the ball to LE (instead of FB), who immediately darts through the opening between RG and C, followed by FB and LT.

QB helps C block his man and force him to the left.

The rest of the wedge plunges in behind RT when the ball is snapped, as before.

See NOTE on diagram sixty-eight.

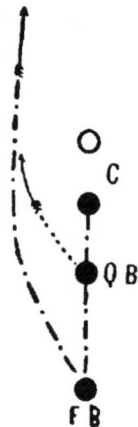
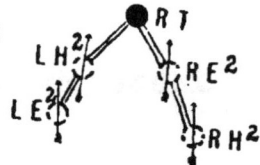

RH

FB

Diag.68

RE

QB
RT
LT

RG
LG

C
QB

FB

LG
RG

LH
LT

RT
LH²
LE²
RE²
RH²

68. Feint wedge and full-back between left guard and center.

After the plays shown in diagram sixty-six and diagram sixty-seven have been worked a few times in succession, RG and RT on the opposing side may find that they can accomplish nothing, as the wedge is upon the other side of the center, and run around in order to help block it. In that case the wedge will form at the signal, and immediately dash in behind RT as before. QB passes FB the ball the instant it is snapped. But FB, instead of plunging in with the ball behind the wedge (as shown in diagram sixty-six), darts through the line between C and LG.

C lifts his man back and to the *right*. LG forces the next opponent in the line to the *left*. QB follows directly behind FB.

NOTE. In case the opposing LG runs around to block the wedge, FB should pass through between LG and C. If RT goes, or RT and QB, he should pass between LG and LT.

NOTE. A strong sequence of movement in the series is as follows :

(1) Play as in diagram sixty-six ; (2) If LH, RH, and QB mass in front of the wedge, play as in diagram sixty-seven ; (3) If they mass close in behind the center play as shown in diagram sixty-nine ; (4) If the opposing RG or RT runs around to block the wedge, play as in diagram sixty-eight ; (5) At all other times play as shown in diagram sixty-six.

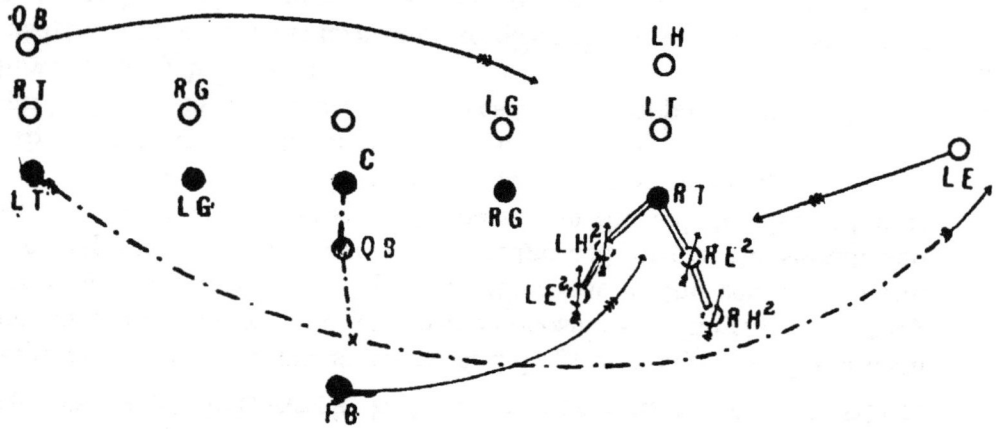

Diag. 69

69. Feint wedge and tackle around the end.

At the signal the wedge is instantly formed as shown in diagram sixty-six, and as the men reach their formation, c snaps the ball. As the ball comes back, FB dashes in behind the wedge in the same manner as shown in diagram sixty-six, when carrying the ball, and the entire formation plunges forward behind RT.

LT leaves his position in the line the *instant* the ball is put in play, receives the ball at x from QB as he passes, but instead of following behind FB, as in diagram sixty-six, swings slightly out around the wedge, in the line indicated, at utmost speed.

The opposing LE may very likely be deceived into thinking the play shown in diagram sixty-six is being attempted, and dive into the wedge where he may be re-pinned by RH.

If the opposing LE does *not* dive into the wedge, RH and RE should dash away to the right ahead of LT the instant he reaches them, to interfere. RH should run directly for the opposing LE, while RE takes the first free man outside of RT.

See NOTE, diagram sixty-eight.

INDEX OF PLAYS.

FOURTH SERIES.

(13) No. 1 and No. 2. End between center and opposite guard.
(14) No. 3 and No. 4. End between the opposite guard and tackle.
(15) No. 5 and No. 6. End between the opposite end and tackle.
(16) No. 7 and No. 8. End around the opposite end.

FIFTH SERIES.

(17) No. 1 and No. 2. Tackle between center and opposite guard.
(18) No. 3 and No. 4. Tackle between the opposite guard and tackle.
(19) No. 5 and No. 6. Tackle between the opposite tackle and end.
(20) No. 7 and No. 8. Tackle around the opposite end.

SIXTH SERIES.

(21) No. 1 and No. 2. Guard between the opposite guard and center.
(22) No. 3 and No. 4. Guard between the opposite guard and tackle.
(23) No. 5 and No. 6. Guard between the opposite tackle and end.
(24) No. 7 and No. 8. Guard around the opposite end.

SEVENTH SERIES.

(25) No. 1 and No. 2. Criss-cross half-back play around the end.
(26) No. 3 and No. 4. Ends criss-cross with half-back in play around the end.
(27) No. 5 and No. 6. Tackle criss-cross with half-back in play around the end.
(28) No. 7 and No. 8. Guard criss-cross with half-back in play around the end.

EIGHTH SERIES.

(29) No. 1 and No. 2.   Half-back criss-cross with the end in play around the opposite end.

(30) No. 3 and No. 4.   Full-back criss-cross with the end in play around the opposite end.

(31) No. 5 and No. 6.   Ends criss-cross and play around the end.

(32) No. 7 and No. 8.   Tackle criss-cross with the end in play around the opposite end.

(33) No. 9 and No. 10.   Guard criss-cross with the end in play around the opposite end.

(34) No. 11 and No. 12.   Tackle criss-cross with the tackle in play around the opposite end.

NINTH SERIES.

(35) No. 1 and No. 2.   Double pass from end to full-back in play around the end.

(36) No. 3 and No. 4.   Double pass from tackle to full-back in play around the end.

(37) No. 5 and No. 6.   Double pass from guard to half-back in play around the end.

TENTH SERIES.

(38) No. 1.   Slow mass wedge from a down.

(39) No. 2 and No. 3.   Feint run around the end from wedge in the line.

(40) No. 4 and No. 5.   Revolving wedge from a down.

ELEVENTH SERIES.

(41) No. 1 and No. 2. Lifting wedge through the center.
(42) No. 3 and No. 4. Wedge from the center around the tackle.
(43) No. 5 and No. 6. Quarter-back around the end from behind the wedge.
(44) No. 7 and No. 8. Feint play from the wedge.
(45) No. 9 and No. 10. Wedge at the end of the line.
(46) No. 11 and No. 12. Play around the opposite end from the wedge at the end.

TWELFTH SERIES.

(47) No. 1. Running mass wedge through the center.
(48) No. 2 and No. 3. Running mass wedge between guard and center.
(49) No. 4 and No. 5. Running mass wedge directly at the guard.
(50) No. 6 and No. 7. Running mass wedge directly at the tackle.

THIRTEENTH SERIES.

(51) No. 1 and No. 2. Free opening play from the center.
(52) No. 3 and No. 4. Double pass opening play from the center.
(53) No. 5 and No. 6. Opening play from the center with team divided.

FOURTEENTH SERIES.

(54) No. 1. Princeton opening wedge from the center of the field.
(55) No. 2. Yale modification of Princeton wedge.

(56) No. 3.　Princeton split wedge.
(57) No. 4.　Yale split wedge.
(58) No. 5 and No. 6.　Side play from wedge at center.

MISCELLANEOUS.

(59) No. 1..　Hard running wedge with loose formation.
(60) No. 2.　Harvard flying wedge.
(61) No 3.　Guard drops back and bucks the center.
(62) No. 4.　Guard through his opening on the same side. .
(63) No. 5.　Full-back feints a kick and runs around the end.
(64) No. 6.　Full-back feints a kick and half-back darts through the line.
(65) No. 7.　Criss-cross play from the side line.

FIFTEENTH SERIES.

(66) No. 1 and No. 2.　Harvard line wedge.
(67) No. 3 and No. 4.　Feint wedge and end between guard and center.
(68) No. 5 and No. 6.　Feint wedge and full-back between opposite guard and center.
(69) No. 7 and No. 8.　Feint wedge and tackle around the end.

TEAM PLAY.

American football is pre-eminently a game for the practice and display of what is known as " team play." No other game can compare with it in this particular. Not that the individual element in skill, in physical capacities, in strategy, and headwork are overlooked, but these are made subservient to the intent of the particular play in hand, and so adjusted to that play as shall best contribute to its success. To get eleven men to use their individual strength, agility, and speed, their wit, judgment, and courage, first in individual capacity, then working with one or two companion players, then as eleven men working as one, is a magnificent feat in organization and generalship.

The individual element, perhaps, is most prominently set forth in defensive play, although there is abundant opportunity in offensive play also for it to show itself; but individual and team play are so closely joined, as a rule, that the beauty of the latter is heightened as the individual efforts of each player are perceived. In defensive work the players have more reason to feel their individuality, because they are often compelled to combat alone one or more opponents before they can get an opportunity to tackle the runner. The defensive system, however, gives a splendid chance for clever team play in the placing of the players, in the general and particular understanding that certain men shall nearly always go through to tackle behind the line; that certain others shall wait to see where the attack will be made and there hurl themselves against it ; that others shall go through the line, or not, just as it seems wisest at the time; and that still others shall never involve themselves in the scrimmage, but act only when the play has been carried into their territory. Furthermore, there is constant opportunity for the exercise of team play in the working together of certain players of the rush line in defense, and also in the working together of any two or three players at special times; for example, when one or two men sacrifice themselves to clearing away the interferers so that a companion can tackle the runner; when one follows hard after the runner to overtake him, if possible, even after having missed a tackle; or helps check him from further advance when tackled, or endeavors to secure the ball.

In the rush line the center and guards work together in defense, having an understanding with each other and with the player hovering in their rear, whenever it seems best to try to let him through on the opposing quarter-back or full-back, or whenever a special defense for certain plays seems best. Likewise the ends and tackles are closely joined in team play, in that they are the players relied on to stop the end plays and those between tackle and end. The most perfect adjustment and team work is needed in doing this, for they play into each other's hands while, at the same time, they seek to tackle the runner. Similarly, but less closely, do the guards and tackles work together in defense against certain plays.

It is an essential point in the working out of this team play between the different parts of the rush line, that the players study most carefully the positions they should occupy to meet the different kinds of play — how far from each other they should stand for this play, how far for that. In doing this, they must have regard for their own freedom to attack, not allowing themselves to take a position where they can easily be tangled up, nor one in which they can give their opponents an advantage in blocking them. Except on wedge and mass plays, the players in defense should draw their opponents apart sufficiently to give themselves space to break through on either side

The backs supplement the work of the rushers in defensive play, arranging themselves behind the rush line at such distances from each other and from the forwards, as shall give the strongest defense. In that degree in which they make their work strong in team play, will they give the rushers encouragement and support in going through the line. The forwards will thus be enabled to play as a unit, because they know that there is a reserve force directly behind them to lend them assistance and make their play safe.

The backs work together in special defense on a kick, arranging themselves, either one or both, in front of the catcher to protect and encourage him, and to secure the ball, if muffed; or one stands behind to make the play safe, or to receive the ball on a pass from the catcher for a run or kick. The ends sometimes come back with their opponents at such times, to bother them all they can and to be in a position to interfere for the catcher, if he runs. The backs, also sometimes have a chance to help one another out by blocking off opponents, while one of their number makes sure of a rolling ball which, perhaps, has been kicked over the goal line or into touch.

When one side has the ball, it is often possible for the opponents to guess in which direction it will be carried, by the way the half-backs or quarter-back stand; by their

15

unconscious glances in the direction they will take; by certain anticipative movements of the muscles; by false starts before the ball is put into play. Further information is often given by the rushers themselves — often by the rusher who is to carry the ball. Frequently the players who are to make the opening indicate by the way they stand, by shifting their positions after the signal is given, or by certain actions peculiar to them at such times, the general direction of the play, and, perhaps, the exact place at which it is aimed. All this is most valuable information and ought to·be imparted to the rest of the team whenever sufficiently positive to be of service. Indeed, the team play of the future will not be considered satisfactory without a set cf signals being used to spread just such information.

At the same time that it is possible to gather much information of this character from the side with the ball, it must be remembered that shrewd players, knowing how they are watched for these tell-tale signs, have cultivated certain false motions, and are using them as points in strategy to deceive their opponents into expecting a different play from the one which is actually made.

From the foregoing, one draws the lesson to hide the intended play. At least, the play must not be indicated by any of these signs which the green player, and too often the experienced player, shows. Thoughtful self-control in every particular is what each player must cultivate, if he would do the greatest service for his team.

Now and then, also, in offensive play the maneuver resolves itself into a test of individual skill, speed, endurance, and headwork; but this is nearly always the outcome of team play in the first part of the movement. Occasionally a mishap furnishes a player a chance to make a run wholly through his own unaided efforts.

The history of the evolution of the hundred and more plays in American football is the history of the development of a " team " game. The perfecting of this has largely increased the number of combinations now possible and has given a wideness in variety of play, and at the same time a definiteness of action for each play, which makes it possible for every member of the eleven to assist powerfully in its execution. In fact, the execution of the play depends on every player doing his particular work for that play. Hence, the interdependence of the players is very close from the moment the ball is down until the run is made, or until a fair catch or a down by the opponents declares that the ball has been released. It is therefore exceedingly important that the adjustment of every factor in the play be made with perfect skill and in exact sequence, from the beginning till the end. It is most important, however, that the starting of the play be well made, for no amount of cleverness afterward can atone for a bungling start.

Team play from a scrimmage should begin the instant the center receives the ball from the hands of the runner (which should be immediately after he is stopped). Every rusher and back should be in position for the next play, and the signal be given before the runner has had hardly time to rise from the ground. The delay of one man in taking his place might be sufficient to spoil the play, whether that man be a rusher or a player behind the line.

As soon as the ball is in play the rushers must give their united support to the quarter-back and the runner, blocking their opponents, if necessary, long enough for the quarter to pass the ball and the runner to get well started. The center and guards especially must work together to protect the quarter while receiving the ball and passing it, and then all or part of them may move elsewhere to help out in the play, or may stay in their positions to make an opening for the runner. There must be the most united work in these preliminaries to the run. Irregular snapping of the ball, either in direction or in speed, which causes the quarter to fumble or to be delayed in getting it to the runner, a poor pass from the quarter, a muff or fumble by the runner, the letting of an opponent through too soon, are usually sufficient to spoil the play.

The rushers will do well in the preliminaries if the runner succeeds in getting up to the line without encountering an opponent, or in the end plays if he is able to get under good headway. They perhaps need only to make a strong blockade in those parts of the line where the particular play is in greatest danger of being checked, but in order to do this well they must regard each other's position as well as their own, touching elbows when necessary, or separating according to the line tactics deemed most effective at the time.

The work of a part of the rushers consists in preceding the runner whenever possible, working together by strategy and combination to make an opening for him and his interferers to go through. The others follow closely from behind to render what assistance they are able. This work comprises the hardest part of the whole play, for it must be executed in the face of the strongest part of the resistance. The rushers can block their men for a second or two, but to block them from three to six seconds is impossible against good players. It is here that the interferers come into especial prominence and value, for they are to clear the way of these free opponents. It is in anticipating the probable positions of the opponents in the vital stage of every maneuver, and in providing the cleverest team play to meet each contingency, that a team excels in advancing the ball by running.

Several things are especially necessary to produce skill. ful team play. First there should be a wise selection of players, and they should be placed in their final positions as early in the season as possible. There also should be such judgment in the arrangement of these players for each position as will produce the least friction in working out the plays, and that arrangement will usually be most effective in which there is the least delay and ill adjustment in making the plays quickly. There should be hard, systematic daily practice, backed by a close study of every play by each player in his particular position. The same players should be used together as much as possible, so that they can become thoroughly acquainted with each other's style of play and know each other's weak and strong points. In this way only can the fine adjustments and combinations which go to make up team play be brought about.

Team play in interference can only be the result of a carefully-planned system in which every player studies the general directions laid down for each play with a view to perfecting his particular work, varying his position on the field whenever necessary, starting like a flash in this play and delaying somewhat in that, blocking his man in one game perhaps in a certain way and in the next in one entirely different, because his opponent plays differently, sometimes taking another opponent instead of his own, when he sees that he can be of more assistance by so doing, and, in fact, doing whatever will most conduce to the furtherance of the particular play in hand.

In most plays the part which each player shall take in the interference can be laid out very definitely, but in the end play, and plays between end and tackle, only part of the interferers are to take particular men; the rest block off whatever opponents come in their path. It is in this free running that there are frequent chances for the display of fine team play in interference in striking the opponent at the nick of time, in pocketing him, in forcing him in or out as it seems best on the instant (the runner being on the watch for either), and in the runner sometimes slowing up to let an interferer who is close behind go ahead to take the man. Very often the reason that a play is not successful is because the interferer is too far in advance of the runner to be of any service to him. Interference must be timely to be effective. It must be the projecting of a helper at the moment a point of difficulty arises — the swinging into line of a series of helpers in timely sequence as the runner advances. Nor must the runner be delayed by the interferers except, perhaps, when the guard comes around on an end play where it is necessary to slow up a little at a certain point to let the guard in ahead.

The execution of nearly all the plays depends for its success on each player doing his duty at the right moment. Here and there in certain parts of the play one or more players must delay a particular work as much as possible, otherwise their action would be immature and so valueless; but for the most part, the movement of each player should be quick and definite, and those plays are most effectively made in which every player does his duty quickly.

Naturally, the end plays and the plays between end and tackle require more delicate adjustment of the players in the interference than do the center plays. In the latter, the interference nearly always must be done after the line has been reached and penetrated. Here the extra men, who rush to the opening as soon as they see where it is, will be encountered, while in the end runs an opponent is likely to show himself here and there and everywhere before the runner reaches the line.

In all mass and wedge plays where the pressure is brought to bear on one point in the line, the team play is not nearly so delicate and skillful. The virtue in the wedge play, be it quick or slow, lies in the power to project great weight and strength on a given point, while at the same time closely protecting the runner.

Every play should be made as safe as possible by having at least one player always in a position to get a fumbled ball, or in case an opponent secured the ball, to prevent him from making a run. Where there are so many parts to every play in snapping, handling, passing, and catching the ball, there is constant danger of a slip. The value of having one or more players behind the runner is frequently demonstrated also, when, by the aid of a timely push, the runner is able to break loose from the grasp of some tackler who has not secured a strong hold on him, and so adds several yards to his run.

In running down the field on a kick the rushers should run in parallel lines two or three yards apart, for most of the distance, converging as they approach the man with the ball, in order to pocket him. The ends approach the catcher in such a way that he will be forced to run in towards the approaching rushers, if he runs at all. All must be on the watch to thwart a pass to another man.

There is a nice point in judgment to be considered by the rushers in going down on a kick. The end men being so far away from where the full-back will stand when about to kick, can start instantly down the field, leaving the half-backs to block off their men if they come through too fast; for the ends' first duty is to be under the ball when it falls. Occasionally, when kicking from near the side line, it may be necessary for the end next the side line to block his man or to push him back as he breaks through to go down the field. What the ends will do in this case, the tacklers should do nearly every time that a kick is made. Both tacklers should feel it their bounden duty to support the ends by going hard after them the instant they judge their opponents cannot reach the full-back in time to interfere with his kick. Hence, any tactics which they can put into practice which will enable them to block their opponents and, at the same time, not delay them in going down the field are the ones to be used. The tackles must bear in mind that the distance from their positions to the full-back is not very great, especially on the side on which the full-back kicks; but while this makes the duty of blocking on that side greater, the other tackle can afford to take an extra fraction of a second from blocking his opponent and use it in a quicker start.

On the guards and center rests the greatest burden in blocking their opponents on a kick; for while there is not that openness in the line, as at the tackle and end, which will let an opponent through quickly, the distance to the full-back is here the shortest and it is usually here that tricks are worked by which one or two men are let through, one usually being the quarter-back. They must, therefore, be very careful not to be over hurried in going down the field, remembering that it is their first duty to block, following the tackles and ends as soon as possible. If the guards and center are very skillful there need be no great delay in doing this, for it is necessary to check their opponents only long enough to enable the full-back to punt over their heads. Whenever it is possible for the guards and center to carry their men before them for a few feet, it is generally safe to leave them and go down the field at full speed It is comparatively easy for the center to do this at the instant that he snaps the ball. Generally there is too much blocking done and too little "following the ball."

In this connection, as a help to the rushers, several points must be borne in mind by the full-back in kicking. It is not enough for him to kick the ball as hard as he can each time it is sent back for that purpose. That would be a poor performance of his duties. He must kick for his team's advantage always, and therefore must regulate the distance, and direct his kick with the utmost skill. Even long and puzzling kicks are dangerous unless closely followed up by the rushers; for, let a good dodging half-back get free, with one or two interferers in a broken field of opponents, and he will be almost sure of a long run.

The full-back must take into account the ability of the rushers to get down the field in time to prevent a run or a return kick and punt accordingly. He may find it necessary to elevate the angle of his kick so that it will give his men time to get under it, or he may find it best to direct the ball straight ahead, in order to give his rushers the shortest distance to run, and at the same time be able to advance in the best formation for checking a run. At least, he must punt the ball where it shall be difficult for the backs to reach it quickly, and so give the rushers the advantage of a longer time to get under it. Especially must he be very careful not to kick the ball diagonally across the field without weighing well the risk involved, in comparison with the chances for increased advantage; for the risks are unusually large in such a kick. It would be well for the full-back to give the rushers a signal as to the direction he meant to kick. This should always be done when he intends to kick off to one side of the field, or when he purposes making a high kick or one outside of bound in order to put his men on side by running forward. The rushers would be able to work some splendid team-play on such occasions.

The question of when to make a fair catch and when to run is well worth the consideration of the backs, who are the ones almost always called upon to exercise their judgment on this point. It was formerly judged best, in handling a kicked ball, to make a fair catch on all occasions. To-day there is a division of opinion, some adhering to the old way, while others prefer to run whenever they get a chance.

There are two points to be considered in deciding this question: First, whether it is possible to kick a goal from the place where the ball will fall, or whether a punt from that point would be desirable; second, whether it will add much to the risk of not catching the ball, if the attempt is made to run. It is clear, that when near enough to the opponent's goal to try a place kick, every effort should be made to secure a fair catch.

When a goal from the field would be impossible, it is almost invariably best to run with the ball, unless this would add greatly to the danger of muffing it. Catching the ball necessitates a positive loss of ground before again putting it in play, and it is doubtful whether this loss is compensated by the advantage of putting it in play unmolested by opponents and behind the whole team under slight headway.

In attempting to run the player will at the worst be forced to make a down, which would furnish only slightly less advantage than a fair catch, while on the other hand it presents opportunities for gain.

FIELD TACTICS.

Clever tactics on the football field depend first of all on the captain's possessing an accurate knowledge of the strength and weakness of his team, both in individual play and in team play. This can all be acquired during practice by carefully noting every play which is made, and giving thought to the strength of the individual men and the value of the play in its relation to the others, both in regard to the perfection of execution and in intrinsic merit from a strategic point of view. It also depends upon the captain's observing as soon as he enters the field and throughout the game, the incidents of the day ; the direction and force of the wind ; the position of the sun ; and the condition of every part of the field. All these points are of great importance in good generalship. Lastly, it depends upon the study which he makes of the way the opponents arrange themselves on the defense, as well as the style of their play when in possession of the ball. He must also seek to find out by trial which of his plays can be used most effectively.

Having the knowledge of the first and second requisites for good generalship, the captain must immediately proceed to find out the weakness and strength of the opponent's defense, not by trying each play in turn and just noting its success, but by using the best tactics the occasion demands, and closely-observing the result on each play. Every play known to be strong because of the ability to concentrate or mass the players at some part of the line, or for any other reason, should be tried at least two or three times early in the game in order to give it a fair test, that the captain may know which will be his most effective plays. It is a mistake to keep pounding away on two or three plays which give an advance of a few yards, just on that account, until after other reliable plays have been given a fair trial. In making this trial, the time should be well chosen, both as to position on the field and as to the number of the down, and the previous loss or gain, if it is the second or third down. It often happens that a powerful play is discarded because in one or two trials it did not work well. The difficulty may have been in its imperfect execution, or in a neglect of duty on the part of one man even, or it might result from the inability of one player to do his work because of circumstances or tactics on the part of his opponents which he could not overcome, but which, later on, he would discover a way to meet.

By confining the tactics to a few plays which have proved successful for more or less gain, the captain limits his play very decidedly and clearly indicates his policy, thereby giving his opponents a knowledge which is invaluable in thwarting him. The result will be that all the available players upon the opposing team will be called from the appointed positions where they had been placed in order to meet the most varied style of plays, and stationed where they can render these particular plays most ineffective. The knowledge that the play will probably be one of a few, also gives every player on the defense a certainty of action which will make his opposition very much stronger. The uncertainty which comes from combating a variety of tactics weakens each man's defense considerably, and puts him at his wit's end to discover what the play will be and how to meet it. It also makes him more liable to be blocked off and pocketed.

Sometimes, to be sure, it is fine strategy to keep pounding away at some particular point or points in the line, in order to draw the attention wholly to this place and to draw the men away from other parts of the line in order to weaken it for a sudden attack ; but this is quite different from the limited style of play so often used, and really, if well done, is a mark of clever generalship.

The captain sometimes uses all his plays in succession simply because he has been accustomed to run through them in practice. This is poor tactics. If it has once been clearly proven that a certain play cannot for any reason be made, every clear-headed captain will realize that it is very poor policy to waste downs in the effort.

A similar mistake sometimes grows out of giving the signals in practice. If the captain or quarter-back in giving the signal is not careful, he will get into the way of unconsciously arranging the plays according to the law of association of ideas, one play following another in unvarying sequence. The principle of sequence in plays would not be fatal, and, indeed, would sometimes be very effective, if the plays are well selected. But account should be taken of the physical capacity of the players ; the duties which they have just been called on to perform ; and the right time and place on the field, in reference to the side lines and nearness to the goal. The great advantage to be gained lies in having the sequence come in the form of a series which is perfectly learned, so that play after play shall be made in rapid succession. The series, however, should not consist of more than from four to six plays, as contingencies often arise which seriously injure their effectiveness. In any case the series ought to be stopped if for any reason it is unwise to make the next play, or if the conditions allow a much better move. A simple signal will indicate that the series is to be stopped. The great virtue in series plays lies in the fact that a certain signal starts the series and each play can be made in the quickest manner, because the players all know what is coming next and are ready the instant the ball is in the center-rusher's hands. Series plays are especially effective against a team which is slow in lining up. They are very valuable also in their moral effect, because of the rapidity and enthusiasm with which the plays are made.

Under a varied style of play where many movements are well executed, the opposing team must exercise the greatest headwork and caution in its defense. If the other team has not already indicated its policy by clearly defining its plays, every one on the opposing eleven will be conscious of so much uncertainty as to what the play will be, that his attack through the line is likely to be cautious and therefore not strong ; or else it is likely to be sufficiently daring to give the opponents a decided advantage in making their plays. When undue caution is exercised on the defense, its effect often is to make the players hesitating. This, when extending throughout the rush line is fatal to a strong defensive game. A daring, reckless defense is far more effective than the cautious defense which makes a rush line hesitate, because of the moral effect on the other team, if for no other reason.

And this leads us to consider the moral effect of certain tactics. The three most effective styles of plays when successfully used are : a kicking game when there are weak catchers behind the opposing line (or when the latter are poorly positioned); end plays ; and dashes through the center in mass or quick wedge plays. These three plays, in the order named, have the most disheartening effect on the opposing team, when the side having the ball has a long, accurate, and scientific kicker who is able to place his punts well, and also to regulate the height and twist which the ball shall take.

Every football player knows the chances for a fatal misplay which hang on a kicked ball : first, because of the difficulty of judging it accurately if it be twisting in certain ways ; second, because of its exceeding susceptibility to currents of air which make its gyrations and deviations excessively perplexing ; third, because of the nicety of final judgment required, even when the player is well under the ball, since its shape and elasticity make it necessary to allow for its full length and its smallest dimension at the same time, also for a quick rebound from the arm or hands. The catcher must attend to all this in the face of a fierce line of rushers coming down on him at full speed, eager to tackle him or to seize the ball if he muffs or fumbles it.

The moral effect of having uncertain catchers behind the line is very telling on the team. If all the hard, wearying work of the rushers and half-backs to advance the ball forty or fifty yards is to be spoiled over and over by muffed punts, even though the ball is not lost to the other side (as it is likely sometimes to be in such cases), there is sure to be a diminution in effort in a short time on the part of the whole team. This comes imperceptibly at first, but comes just as surely, and ere long evinces itself in the more determined and successful efforts of the other team.

Almost equally disheartening, if not fully so, is it to have runs made repeatedly around the ends; because the runs in that locality, if successful, are usually for long gains often resulting in touch-downs, and they arouse the greatest fears in the minds of all the players from a feeling of inability to stop them.   The result is that every effort is centered on anticipating these end plays, and the rushers, instead of going through the line, wait to see if it is an end play, in which case they run out to the side to stop it. That very moment in which there is a hesitancy on the part of the guards and tackles in going through the line, is a moment of triumph for the team with the ball; for it immediately gives them a decided advantage, in that, while perhaps unable before to make progress through the center part of the line, they will now have two strong points of attack. The chances now are that the defense will grow weaker and weaker as the game advances, for unless the end runs are well stopped the players will decrease their efforts somewhat and the tackling will become less and less daring and effective.

It is hard to say which of these two styles of play really has the more discouraging effect on the opposite team. If the eleven which has the poor catchers back of their forwards are successful in making advance by rushing the ball, they have a vast deal to encourage them, even though now and then they lose it all through the muffing of their backs. The period in which they have the ball is one in which their minds are not conscious of the weakness of their own defense but are completely taken up with the good work they are doing, and they are unanimous and bouyant in it. That period of success does much to keep up their spirits during the time when the other side has the ball and their fears are so all-powerful.

When a team is able to make frequent runs around the opponent's end, there is perhaps less to actually dishearten them than in the preceding case, for there is less fear of losing the ball. It can be gotten only through a failure to advance the five yards in its three trials; through a fumble ; through a penalty imposed by the umpire; or through a kick. The latter will be tried probably only under extreme conditions where there has been a loss of yards, while in the kicking game mentioned above, the side not in possession of the ball always has the hope of securing it.

That captain is not a good general who follows out the same tactics in each game; who, having perhaps worked out a system of plays which his men could best execute, attempts to apply this system in every game, regardless of the composition of the opposing eleven and their systems of defense and offense. The captain, in truth, has learned a good deal when he has learned what plays his team can best execute, and he has most valuable, though far from complete, information for conducting a wise campaign against the opposing eleven. He still has much need to exercise his generalship as to whether this point of attack should be assailed three or fifteen times; this place a few times; and this place not at all, or perhaps only once or twice for the sake of trial or strategy.

Oftentimes, the rusher can give invaluable information to the captain as to his own ability to handle his opponent, where for example the opponent so places himself constantly as to render it an easy matter to get him out of the way for certain plays, although it is impossible to

16

move him on other plays. This is especially true in handling a large man who stands constantly in the same way; as for instance, well over to the side of his opponent. It would be comparatively easy to block such a man for opening up a hole in one direction, but almost impossible to shove him in the opposite way. Such information would furnish the captain valuable data on which to base certain tactics, and would inform him that he could doubtless make plays to one side of this man and seldom if ever on the other side.

It would be foolish, even if it were possible, to lay down a complete system of tactics which should be followed in a game. Indeed, the wonderful part of football is, that it is a game which cannot be worked out by rule and learned by note. One play does not follow another in sequence, but only as the captain or commander of the day directs.

What makes the game preëminently one requiring science and brains, is that to be well played the captain must use the utmost wisdom and strategy in directing the plays, and the players to a man must do their duty in executing them. Very many points of advantage and disadvantage must constantly be borne in mind, or else the best generalship and results cannot follow. It is far from true to say that the captain must simply take into account the strong and weak points of his opponent's play, together with the incidents of the day and field, such as the direction of the sun and the condition of the grounds in each particular part of the field; he must also have regard for his men, selecting his plays with such wisdom as to secure the greatest economy of physical energy with the greatest result, so that no man nor men shall be overworked at any time of the game and thus be incapacitated.

No captain is a good general who does not know the limitations in strength of his ground-gainers, and who does not take this into account in directing the play. Men differ greatly in their power to repeat a performance quickly; essentially, then, in their powers of endurance. Some men can do effective work only when in first-class condition; that is, when they have had a certain length of time to recover after each effort, they can be relied on for a good gain, if not a brilliant run. Then, there is a vast difference in the kind of play as to the drain on a man's strength. End runs, and runs in which a considerable distance is covered, or runs in which there is a good deal of dodging and struggling to get loose from tacklers, are the most taxing on the wind and strength. Most men can stand two or more dashes through the line in quick succession, or two or more mass and wedge plays where the runner does not run fast for a long distance before being tackled. But when a run has been made which has called for a vast deal of energy the captain should not fail to notice it, and in calling the next two or three plays, choose such as do not ask for too much strength from this player. The star runner as a rule is the one who suffers most from overwork through injudicious leadership.

This does not preclude the fact that there are occasions in the game when some player or players must be forced to draw heavily on a reserve fund of energy in order to secure a permanent advantage or to prevent disaster. It sometimes seems necessary when nearing the opponent's goal, that some player be put to his supreme test of strength in order to secure points, and likewise, when it is necessary to carry the ball away from one's own goal, and there is only one man who is sure to meet the crisis ; but these are in truth critical periods and are exceptions not to be mentioned in this connection.

We know that it is sometimes considered clever tactics, when there are strong substitute players for certain positions, to work men in these positions to their utmost limit of service, and then "have them get hurt" in order to substitute a fresh man or men. If this be shrewd, it is at least not honest tactics.

If a team is not capable of playing an uphill game, or is one which is strongly affected by success and repulse; or, if the opposing eleven is one which is similarly influenced, the tactics should be those most likely to produce the exultation of success on the one hand, and the feeling of discouragement on the other. The plays should be those which can be executed quickly, and which have a certainty of gain with little risk of loss ; which combine the efforts of every man in the eleven sufficiently to make him feel that he has an important part in them; which bring the energies of the opposing eleven, particularly the rushers, to the severest test, taxing especially the wind and courage.

It must always be remembered, as a point in tactics, that the side in possession of the ball has a great advantage, especially if the other side is weak in defensive play, and that it requires a greater outlay in strength and wind to check plays than it does to make them. It is likewise true that the courage of a team may be measured by its promptness and determination in defense. If a team repeatedly and continuously comes up to the scrimmage, after being outwitted and outplayed, it has the true courage, the courage which would probably enable them to win if possessed of an equal degree of skill in team-play.

What style of game shall a team play? That depends on many contingencies. Setting aside for the time the incidents of the day, such as wind, rain, and sunlight; the soft, slippery, and rough places in the grounds; the up and down grades ; — not even taking into account the strength and weakness of the opponents, and the contingencies which arise, let us consider solely the composition of the team, and see if we can deduce any style of play which applies to teams made up of certain types of men.

Without defining the make-up of the team, except on general terms, we see that when the rush line is strong and heavy, the chances are that they will be able to handle their opponents and make good openings for the dashes through the line. Plunges through the central part of the line will probably be the most effective, if the center guards and tackles are large and strong men. If the backs are slow and heavy also, a center game will probably be the only kind they can play with success. And the result is that this will be the style of game adopted; not perhaps because the captain has analyzed the reasons for the ability of the backs to make advance in that place, and their inability to circle the ends, for example; but just because that is the part of the line in which they can make their gains every time. Perhaps it will occur to him that those same backs can be so quickened in starting and running, and then so well guarded, that they will be able now and then to try an end play, or a tackle and end play successfully, and by so doing, strengthen that very center play. The chance for making a successful end play is increased where a center game is being played, because the ends will be likely to draw in somewhat to help the center.

When the center men of the line are rather light, if the backs are heavy and slow, the advantage will still be in attacking the openings between the center and guards and between the guards and tackles; for, if the backs and ends mass on these places, as they can do quickly and powerfully, they can still force a few yards at a time, and now and then break through for considerable gain. When well massed, this can be played even against the strongest centers. All that the rush line will need to do is to hold their men momentarily until the backs get under headway, and the combination of so much weight and power will be sure to make advance when well directed. If it be remembered that the advantage is always with the side which has the ball, and if the players, though checked now and then, go into each play with undaunted courage, advance will surely be made.

As a general rule, when a team has light, swift runners behind the line, they should lay the emphasis on plays around the end and between the ends and tackles. Not that they should confine themselves to those points of attack, but it would be foolish for a team composed of such material not to perfect the plays in these parts of the line, because of the ability of the backs to move quickly to these remoter places. Such men, too, are not so well built for the hard, plunging work in the center, and will probably stand less of it, and be less effective, than heavier backs. This of course depends in part on the build of the men, but in general it is true.

But even if the backs are equally good in plunging into the line, it would be better policy to keep the line spread out, for no runner can make much gain through a close line. Swift drives through the line can be made frequently, and are usually very telling when the line, being spread out, is opened up for these little backs to come darting through. But if the backs and the central part of the rush line are both light, while those of the opponents are heavy, the end style of play must of necessity be depended on, or the opposing rushers will be able to resist the plunges. Furthermore, it will be exceedingly hard to make holes through the line, and, in fact, even to hold their opponents long enough for the backs to get up to the line.

The question of what shall be the proportion of end plays and plays between the ends and tackle, to the plays through the other four openings in the line depends, of course, very largely on the backs. The composition of the rush line as to strength and skill, especially the center, guards, and tackles, also affects the proportion.

On the ordinary college and preparatory school team, the relative effectiveness of an end game to a center game would be much smaller than where the teams are better trained, simply because the risks are larger; for, while the defense against well executed interference would be much weaker, the attack also is much weaker.

Every end play and play between the tackles and end is attempted with a much greater risk from actual loss of ground, or with a loss of a down with no gain, than are the plays in the center. The reason is that the rushers are given time to break through the line while the runner is moving out to the point of attack, and unless well protected he will not reach the opening.

Further, this movement for a considerable distance is almost entirely sidewise before an advance can be made, while in the plays in the central part of the line the rushes are made nearly straight forward, except when the rushers take the ball, and the runners scarcely ever fail to reach the line. The times when there is no gain whatever and when there is an actual loss are comparatively few, for the runner, catching the ball at full speed, is up to the line in an instant, and then it becomes a question how far he can advance beyond that point. Taking these elements of risk into account, it would seem that the proportion of plays at the end to plays through the line should not be larger than one to three, and oftentimes less, even where a team is able to use both styles effectively. The only occasion for a larger use of end plays than this would be when the runner seldom fails to reach the line, and is usually good for a gain. In that event the large element of risk has been taken away, and the proportion of use should then depend on the relative amount of gain which the trials have shown can be secured from each with the least expenditure of energy.

Right here it might be well to add that it requires more skillful generalship to know when to use an end play than when to make a play through the center. It is only occasionally that the ball is down so close to the side lines that all four openings in the center are not available on account of running outside the line, while it is frequently the case that the ball is down near enough to the side line to limit the end play to one side, that is, to two openings. Nor is this enlarged space on one side of the field sufficient compensation for the loss of the two points of attack, but it adds to the science of the game, as it requires more varied tactics and maneuvers.

It is poor tactics to keep trying end plays when it has been clearly proven that it is not possible to make them and that there is a likelihood of a loss in the trial. If it seems best to try the end for the sake of keeping the opposing line spread out so that the center plays can be made more successfully, the most propitious times should be selected. It should never be on the second or third down, because the risk of losing the ball by failure to gain the requisite five yards would be entirely too great.

There are times when an end play should not be used at all, or very rarely, on account of the risk involved; as, for example, when the ball is being carried out from under the goal where it has been forced by the opponents. Anywhere within the fifteen or twenty yard line it is much better to trust to bringing it slowly out a few yards or feet at a time, sufficient, perhaps, to secure only the requisite five yards in three trials. Beyond the twenty-yard line and up to the thirty-five-yard an end play should be tried only on the first down, or, in rare instances, on the second down, unless the risk of losing ground, and subsequently the ball, is worth taking. In such cases the possession of a powerful punter behind the line, who could place the ball well out of dangerous territory if necessary, might be a sufficient reason for attempting a long kick down the field. It does not seem, however, that it is necessary to run any risk of losing the ball if there is good reason for not playing a kicking game, for there will be ample chance to try an end play on the first down. Mistakes in generalship are frequently made right along this line in nearly every game which is played, an end run being sometimes tried on the third down when there is less than a yard to gain. Better gain the yard or two by the surest ground-gaining play and then try an end run on the very next.

When inside the opponents' twenty-five-yard line the greatest skill must also be used, and the aim should be to get the requisite five yards by the most reliable tactics. Plays which risk the loss of ground and the ball should be sparingly used, and every caution and strategy be exercised to place the ball across the line. Nor should there be less prudence because a team has a good drop kicker. The proportion of goals secured from drop kicks is not more than one in every four or five attempts, with the best kickers in America, and the most certain way to score will be to strain every nerve to place the ball across the line by steadfastly holding the ball and using the drop kick only as a last resource.

Every now and then a point is lost unnecessarily when the ball is in the possession of a team under its own goal. It is judged not wise to kick. Perhaps the wind is strong in the opposite direction and there is no reliable punter, or perhaps it would simply give the opponents a fair catch from which to make a try for goal if kicked. The captain also realizes that if the opponents secure the ball they will force it over. Two downs may already have been used up and ground lost in vain attempts to advance the ball by running. There seems to be no other alternative, and so another trial is made, but without avail, whereupon the ball goes to the other side. Under these circumstances it would be well for the captain to remember that by making a safety touchdown and allowing the opponents to score two, he could have brought the ball out to the twenty-five-yard line and prevented a probable six points.

The mistake is often made of frequently using end plays when the ground is slippery and soft from rain. Nothing can be more foolish, unless the aim is to get the ball on firmer ground, for with insecure footing it is impossible to start quickly, run fast, or turn and dodge quickly. This makes it easy, also, for the opposing eleven to stop the runner and nearly always with a loss of ground. The same is true, in a measure, when the ground is soft or very sandy. It is comparatively hard to make end plays even when there are no unfavorable conditions, when the ground is firm and level.

He is a wise general, therefore, who notes the field carefully, knowing where all the soft and slippery and rough places are, as well as where the good ground is, and then keeps them in mind throughout the game, and makes his moves wisely in reference to them. Few captains take the field sufficiently into account in directing the plays, so that the greatest advantage can be secured by avoiding the hindrances as much as possible. Again and again unsuccessful trials to advance have been made in muddy places, when, with one well-planned move, the ball could have been placed on solid ground with little or no sacrifice, and a vast advantage secured. It is usually worth the loss of two or three yards, and oftentimes more, to make an end play in order to give a better footing to the backs and the rushers for putting the ball into play, for handling it, for making holes, and for starting, running, and dodging.

When the ground is very slippery, all plays which cause the runner to move a considerable distance sidewise and across the field before turning to advance, and all plays requiring a sudden change in direction, whether when under strong headway or not, are hard to gain ground on, and, therefore, must be used with great judgment. Equally hard to make are the plays in which the tackle and guard and end carry the ball around for a run through one of the openings on the opposite side of the line. There is not, however, the chance for so much loss of ground in these plays, as usually played, that there is in a run out to the end by the half-backs, because the former run closer to the line and the play is not so quickly perceived.

It naturally follows, then, from what has been said, that those plays which send the runner directly forward;

those in which the impetus from the start is more forward than sidewise; those in which the runner does not have far to run before he strikes the opening; and those in which he can get the greatest protection and assistance quickly, are the plays to be relied on when the ground is soft, sandy, or slippery.

In bringing the ball in from the side lines, the privilege is given of having it down anywhere from five to fifteen yards from that line. This option of ten yards should be valuable in determining the tactics to be used next. Too often is it the habit for the captain to shout out, " Bring it in fifteen," whether the " fifteen" would carry them into a mud hole, or whether there was a positive advantage in operating from a nearer point to the side line by avoiding the usual custom of an end run, and sending the runner through on the other side. Generally the fifteen yard point is the best place to have the ball down, but not always. The ten-yard point has decided advantages in making certain side-line plays, because the opponents will reason that the chances are in favor of an end play being attempted, and will draw one or two men away to strengthen their defense in that quarter. These they will feel that they can well spare from that side without very apparently weakening the defense, because they are protected from long runs by the side line.

The side line does not enter into the consideration in field tactics as much as it should. As a rule, it is considered a misfortune when the ball is down within less than ten yards of this boundary line, because it gives the opponents a good chance to anticipate the play, which is likely to be a run around the other end. The free men who are behind the rushers nearest the side line rarely fail to move over as far as the center-rusher. This leaves the defense of that part wholly to the rushers, supported by the side line, and is the best situation possible for making certain plays. Long runs, however, cannot be expected, and the captain must be contented to work steadily up the field by short gains. After several dashes into the line, of this kind, an end run suddenly carried into execution may have considerable chance for success.

This suggests the thought that it is possible to use the side line helpfully when the ball is down very near it and when it is impossible to make any strong plays because of the limitations which must be met in such a situation. At such a time, instead of attempting to make a run out toward the end, or tackle, which will be expected, the play should often be straight forward or on the side toward the boundary line, until the runner is finally pushed over the line and has the privilege of bringing the ball in to a more favorable position from which to operate.

Furthermore, the position near the side line can be made more useful in working tricks than a point nearer the center of the field, for reasons which are evident.

There is no question that kicking the ball has not entered into the tactics of football as largely as its possibilities would warrant. There are many reasons for this. First, there is only here and there a team which has a reliable kicker. Punting and drop kicking are practiced by a few only, and, for the most part, not intelligently and successfully. It is a science with several points of skill to be acquired. Second, many teams have an uncertain punter who does not himself know exactly where the ball will go, whether far down the field or just over the rush line, along the ground or to one side, and so place such little confidence in the value of kicking under so great a risk that they will usually trust to a run, even on the third down, if the distance which they have to gain is not too great. Third, in all but a few leading colleges when the teams are evenly matched, the question of points is largely a question of which side has the ball. [The offensive game is much better developed than the defensive game, and it is not infrequent for one team to carry the ball from one end of the field to the other without losing it.] Under these circumstances the necessity for kicking is seldom felt, and they would rather take the risk of not gaining the requisite number of yards, than release their right to the ball by an uncertain kick. Fourth, it is a fact that most punters can not kick accurately if forced to punt quickly. They are, therefore, compelled to stand so far back of the rush line that the value of their punt is decreased by several yards, or else they run the risk both of a poor punt and of having it stopped by the opposing rushers who break through the line.

No better proof of the value of a good punter behind the line is needed, than to see a game in which one side is visibly weaker than the other in its power to advance the ball by running, but which, possessing a strong punter, is able to keep its opponents in check. Frequent punts are doubly effective when the opposite side is without a good kicker, or is not accustomed to a kicking game.

The worth of an accurate kicker is magnified very much when there is a wind in his favor. Comparatively few games are played without a wind to help or interfere, according as it is favorable to one side or the other. When the wind is in the favor of one side, they should be able to use it to the greatest advantage. The captain should be alive to its value, and make it a powerful factor in his tactics. It would then be a question whether it would not be wise to kick the ball just as soon as it was secured, provided, of course, it was not so near the opponent's goal that it would be wiser to hold the ball and attempt to rush it over. Certain it is that a side should never fail to kick on the third down except on account of the liability of kicking the ball over the goal line when inside of the twenty-five yard line, or because so close to the goal line that it is worth taking the risk of losing the ball in making a supreme effort to get it over.

When there is danger of the ball being kicked across the goal line a clever punter will usually aim to kick the ball across the side line into the touch as near the goal line as possible. This is intentional and is quite different from the juvenile efforts which do not take the wind or position into account when punting from near the side line and send the ball outside, only a few yards away.

It is sometimes good tactics on the third down, when there is considerable doubt whether the required advance can be made, to have the full-back kick the ball across the side line with no intent perhaps of a gain in ground. While giving the opposing team technically an equal chance, it is wholly with the purpose of having the end-rusher secure the ball, which will be upon the first down. The kick must be well placed, of course, and must not be so much forward that there will be great risk of the opponents securing the ball, and also not so far ahead that the full-back cannot put his men on side easily. The end man on that side must also know of the full-back's intention, and place himself well over toward the side line. Such a kick cannot be attempted safely when the full-back is not able to place his punts with great accuracy. The occasions when the use of such tactics would be wise, might be when the side in possession of the ball was able to make good advances by running but had lost ground, perhaps through a misplay ; or when they had the ball inside their opponents' twenty-five yard line and were not in a good position to try a drop kick ; or when the risk of making the required gain by running would be too great.

Right here would come in the question of a drop kick on the third down when inside the twenty-five yard line, and in fair position to make the trial. It is safe to say that, in general on the third down, this should be the play called for. It is for the captain to decide whether the trial is worth the making; whether the nearness and angle to the goal, and the quickness and skill of the kicker warrant a drop kick in preference to the chances of making a further advance by running.

If a run is attempted without gain the ball will be down where it is for the other side. When the kick is made on the other hand, there will be a possibility of having the ball stopped by the opposing rushers, and a run made up the field; or, if the goal is missed, the opposing team will be allowed to bring the ball out to the twenty-five yard line.

The captain must weigh all these possibilities before making his decision.

The great advantage in the wind does not consist alone in the increased distance the ball can be propelled, but also in the increased likelihood that some one upon the side which kicked will again secure the ball on a muff or fumble. The wind has added to the problem of the player who attempts to catch the ball these points of difficulty: greater distance covered by the ball, an increased speed, and a greater probability that the ball will suddenly veer to one side or the other from the line of direction.

The increased advantage of a favoring wind is in direct proportion to the strength of the wind. If the wind is very strong, the side which does not have its assistance is severely handicapped, and for the time is not able to do any effective kicking. Even with the best punters, it is impossible to drive the ball far in the face of a strong wind, and then the kick must be low or the wind is likely to blow it back near the spot from which it was kicked. On the other hand, when kicking for distance with the wind, it is usually better to kick the ball high, in order that the wind may affect it more powerfully during the longer interval of time in rising and falling.

There is also an economic reason for kicking the ball whenever it can be wisely done. It is a good way to rest the backs in order to save them for the supreme effort of carrying the ball across the line; for, if the ball has been carried for a considerable distance, they will be likely to be somewhat fatigued as they approach the goal line, and they will be weakest where and when the opposing side always puts in their most determined and desperate resistance.

It is a severe test of a team's courage to bear up against a kicking game in the face of a strong wind; for, even if they are able to make good gains in return by running, the players are constantly fearing a slip or fumble, which will give the ball back to the other side only to have it returned with all the chances of a misplay, if not a gain in ground. The effect of the wind also is to make the side against it think that they are working very much harder than their opponents just to hold their own.

There is no question as to the value of having every member of the team able to run with the ball when it is possible and wise. The more varied the style of play, provided it is strong, or is likely to be successful because unlooked for, the more powerful would be the plan of attack and the less effective the defense. This is true for two reasons: first, it keeps the opposing team constantly guessing as to what the play will be and enables the side with the ball to secure advantages through the variety of its play; second, it distributes the labor and secures the advantage of fresh strength, while it rests the main ground-gainers. For these reasons, then, it is well worth the while to run the guards, tackles, and ends, although these are not in as advantageous positions for gaining ground as are the half-backs and full-back.

The most valuable of the three rush-line positions for ground gaining is the tackle, because from that position the runner can get under sufficient speed to carry him forward against opposition, and he can also secure the most protection and help. The run also can be made in the quickest time and without being immediately noticed.

The end position, when the end plays behind the line and near the tackle, comes next in value of the line positions for running with the ball, because of the large number of interferers ahead. If rightly played by a fast runner, the end will be able to make good advances between the tackle and end, and even around the end on the other side.

The guard is in the hardest rush-line position for advancing the ball, because it is impossible for him to get under speed when making a quick turn around the quarter-back, and on the other hand he cannot afford to run out to the end, because he would be sure to be tackled whether he ran close to the line with little interference, or ran farther back with better interference but with greater risk of loss of ground.

SIGNALS.

In the modern game of football it is absolutely necessary that before each play a signal should be given, which
will inform every man on the team of the movement
about to be executed. Every player has a special duty
to perform each time the ball is snapped, and unless he is
informed beforehand of the evolution intended, it will be
impossible to render the requisite assistance. It is of
equal importance that the opposing team should be kept
in absolute ignorance in regard to the intention of the
play, so that they may not anticipate and thwart it.

That code of signals will be best, then, which will indicate in the simplest manner the play intended, while at
the same time being unintelligible to opponents. Too
frequently such a complicated system of signals is adopted
that the players themselves become confused, or at least
are unable to comprehend the order upon the instant, and
the momentary delay thus caused proves a great disadvantage. There is far less likelihood that the opposing
team will be informed by the signal what play is intended, than that they shall discover its probable direction by the position assumed or nervousness betrayed by
some one of the backs or rushers.

There are three systems of signals which have a practical value : Sign signals, word signals, and number signals. Sign signals possess one advantage which neither
of the other two can claim. They can be understood
with readiness amid the most deafening cheering from the

side lines. It often happens that the cheering is so continuous at critical moments during the great matches, where many thousand people are assembled, that for several moments the play is almost paralyzed on account of the inability of the captain to make his orders heard. It is readily perceived what an advantage it would be to have a code of signals which would direct the play rapidly and unerringly at such a time.

On the other hand, there is, perhaps, more danger that the opposing team may notice and soon learn to under_stand signs than when spoken signals are used, for it is necessary that each man on the side shall look at the quarter-back or captain at the time when he gives the signal (usually this will be when the men are lining up), and this will of necessity attract more or less attention to what it is expressly desired to cover up. Every team would do well, however, to have a complete system of sign signals, which they can use at critical times in case of emergency.

The following extract from a code once in operation will furnish suggestions which will enable any ingenious captain to devise a practical set: Pull up trousers on right side — RH between C and RG. Pull up trousers on left side — LH between C and LG. Right hand on right thigh — RH between RG and RT. Right hand on left thigh — RH between LG and LT. Right hand on right knee — RH between RT and RE. Right hand on left knee — RH between LT and LE. Right hand on collar on right side — RH around RE. Right hand on collar on left side — RH around LE. Right hand on chin — RT around between LG and LT. Right hand on right hip — RE around the LE. Pull on jacket lacings — kick down the field.

Similar motions with the left hand will direct corresponding plays in the opposite direction. The motions should be made so naturally that they will not attract attention, but in deciding upon movements care should be taken not to select those which will be used involuntarily, lest signals be given sometimes without intention.

In the system of word signals peculiar expressions, such as " Brace up now," " Now brace," " Hold your men hard," " Tear up this line," " We must do better now," and the like, introduced by the captain with a few off-hand sentences before each play, direct the next movement. Again, speaking to the left tackle may indicate that the left half-back is to run around the right end, each man being made to indicate a different evolution; and a word of encouragement or·blame thus be made the signal for the next play.

Perhaps the system of signaling by numbers is most simple and satisfactory, for it admits of a great variety of combinations, and the key will not be readily detected. Sometimes a long sequence of numbers are called out, the signal being conveyed by the first two or three, and the others being added merely to mystify the opposing side, but a combination of three numbers is rather preferable.

A very simple code may be arranged, in which each opening is given a number, and each player a number. The combination of two numbers, then, will indicate the man who is to receive the ball, and the opening through which he is to pass, while a third will be called for the sake of deception. For example: We will suppose that the openings in the line, as they radiate from the center, have been numbered 4, 6, 8, and 10, respectively, upon the right, and 5, 7, 9, and 11 upon the left; the center-rusher will be No. 1, RG will be 2, RT will be 4, RE will be 6, and RH will be 8; while on the left LG will be 3, LT will be 5, LE will be 7, and LH will be 9, with FB 11. We will further suppose that but three numbers are to be given each time; that the first number called will mean nothing; the second number called will indicate the player who is to receive the ball; and the third number the opening through which he is to pass.

To illustrate: The captain calls "9, 5, 8!" The 9 means nothing. The second number indicates the player who is to receive the ball, which in the present instance is No. 5, the left tackle. The third number shows the opening through which he is to pass — in this case No. 8, and hence between RT and LE. The interpretation of the signal, then, is that LT is to receive the ball, pass around the center, and dash into the line between RT and RE.* Thus any combination desired may be effected.

If, after a time, the opposing team discovers the signal for one or more of the plays, the entire system may be changed by simply informing the team by a peculiar signal, previously arranged, that the first number will thereafter indicate the opening, while the third will indicate the player who is to take the ball. The three numbers admit of six different arrangements, and the team should be drilled upon at least three of them until they can execute the plays with equal readiness under each arrangement.

In more difficult systems each play is given a separate number, which number may be called out either first, second, or third, as determined. Again, letting each play be indicated by a particular number, as before, the

---

* See diagram nineteen.

sum of the last two numbers is taken to make the number desired. This latter system, though, perhaps, a little more difficult, will prove the most satisfactory.

If two numbers are to be added together, the captain will do well to make one of them quite small, and call the larger number of the two first, for the addition will be performed by all much quicker and with less effort. During the first of the season it will be well to use one particular number to represent a play, and when these have been thoroughly learned it will be but a comparatively easy matter to change to the sum of any two.

When the number for the play has reached twenty, it may make the signals easier to have all the numbers between twenty and thirty indicate a certain other play; all the numbers between thirty and forty, another; and so on.

As the kick is a frequent play, and as it is nearly always apparent, it may be well to have two numbers, either one of which will be the signal for a kick down the field.

Enough has now been said to suggest how a practical system of signals may be devised.

AXIOMS.

Line up quickly the moment the ball is down and play a dashing game from start to finish.

Never under any circumstances talk about your hurts and bruises. If you are unable to play, or have a severe strain, tell the captain at once. He will always release you.

When thrown hard always get up as if not hurt in the slightest. You will be thrown twice as hard next time if you appear to be easily hurt by a fall.

When coached upon the field never under any circumstances answer back or make any excuses. Do as nearly as possible exactly what you are told.

Always throw your man hard, and toward his own goal, when you tackle him.

Never converse with an opponent during the game, but wait until the game is over for the exchange of civilities.

If you miss a tackle turn right around and follow the man at utmost speed; some one else may block him just long enough for you to catch him from behind.

Never play a "slugging game"; it interferes with good football playing.

Try to make a touch-down during the first two minutes of the game, before the opponents have become fairly waked up.

Play a *fast* game; let one play come after the next in rapid succession without any waits or delays. The more rapidly you play, the more effective it will be. Therefore *line up quickly* and get back in your regular place instantly after making a run.

When thrown, allow yourself to fall limp, with legs straight, and then you will not get hurt. Do not try to save yourself by putting out a hand or arm; it may be sprained or broken. If you are flat on the ground you cannot be hurt, no matter how many pile on top of you.

Always tackle low. The region between the knees and waist is the place to be aimed at. When preparing to tackle, keep your eyes on the runner's hips, for they are the least changeable part of the body.

Lift the runner off his feet and throw him toward his own goal. When not near enough to do this, spring through the air at him and hit him as hard as possible with the shoulder; at the same time grip him with the arms and drag him down. Always put the head down in doing this and throw the weight forward quickly and hard. Crawl up on the runner when he falls and take the ball away if possible; at least prevent its being passed.

When the runner is in a mass, or wedge, drive in and lift his legs out from under him, or fall down in front of him.

If the runner's feet are held, push back on his chest and make him fall toward his own goal.

Don't wait for the runner to meet you; meet the runner.

Always have a hand in the tackle. Don't "think" the runner is stopped; make sure of it.

Follow your own runners hard; you may have a chance to assist him, or block off for him. Always be in readiness to receive the ball from the runner when he is tackled.

Fall on the ball always in a scrimmage, or when surrounded by opponents. When the ball is kicked behind your own goal, or across the side line, do not fall on it until it stops unless there is danger of the opponents being put on side.

Put your head down when going through the line and dive in with your whole weight.

Call "down" loudly, but not until it is impossible to make further advances.

Squeeze the ball tightly when tackled, or when going through the line.

Never under any circumstances give up because the other side seems to be superior. They may weaken at any moment, or a valuable player be ruled off or temporarily disabled. Let each man encourage the others on the team by monosyllables and keep up a "team enthusiasm."

Be the first man down the field on a kick.

Block your men hard when the opponents have the ball.

Tear up the line, break through and stop every kick that is made.

Never take your eyes off the ball after the signal has been given, if you are a man behind the line.

Do not be contented with a superficial reading on football, but *study* it carefully, if you would master it.

RULES ADOPTED

BY THE

AMERICAN INTERCOLLEGIATE FOOTBALL ASSOCIATION

FOR 1893.

RULE 1. (*a*) A drop-kick is made by letting the ball fall from the hands and kicking it at the very instant it rises.

(*b*) A place-kick is made by kicking the ball after it has been placed on the ground.

(*c*) A punt is made by letting the ball fall from the hands and kicking it before it touches the ground.

(*d*) Kick-off is a place-kick from the center of the field of play, and cannot score a goal.

(*e*) Kick-out is a drop-kick, or place-kick, by a player of the side which has touched the ball down in their own goal, or into whose touch-in-goal the ball has gone, and cannot score a goal. (See Rules 32 and 34.)

(*f*) A free-kick is one where the opponents are restrained by rule.

RULE 2. (*a*) In touch means out of bounds.

(*b*) A fair is putting the ball in play from touch.

NOTE. The ball adopted and used by the American Intercollegiate Association is the "Spaulding J." ball.

RULE 3. A foul is any violation of a rule.

RULE 4. (*a*) A touch-down is made when the ball is carried, kicked, or passed across the goal line and there held, either in goal or touch-in-goal. The point where the touch-down scores, however, is not necessarily where the ball is carried across the line, but where the ball is fairly held or called "down."

(*b*) A safety is made when a player guarding his goal receives the ball from a player of his own side, either by a pass, kick, or a snap-back, and then touches it down behind his goal line, or when he himself carries the ball across his own goal line and touches it down, or when he puts the ball into his own touch-in-goal, or when the ball, being kicked by one of his own side, bounds back from an opponent across the goal line and he then touches it down.

(*c*) A touch-back is made when a player touches the ball to the ground behind his own goal, the impetus which sent the ball across the line having been received from an opponent.

RULE 5. A punt-out is a punt made by a player of the side which has made a touch-down in their opponents' goal to another of his own side for a fair catch.

RULE 6. A goal may be obtained by kicking the ball in any way except a punt from the field of play (without touching the ground, or dress, or person of any player after the kick) over the cross-bar or post of opponents' goal.

RULE 7. A scrimmage takes place when the holder of the ball puts it down on the ground, and puts it in play by kicking it or snapping it back.

RULE 8. A fair catch is a catch made direct from a kick by one of the opponents, or from a punt-out by one of the same side, provided the catcher made a mark with his heel at the spot where he has made the catch, and no other of his side touch the ball. If the catcher, after making his mark, be deliberately thrown to the ground by an opponent, he shall be given five yards, unless this carries the ball across the goal line.

RULE 9. Charging is rushing forward to seize the ball or tackle a player.

RULE 10. Interference is using the hands or arms in any way to obstruct or hold a player who has not the ball. This does not apply to the man running with the ball.

RULE 11. The ball is dead:

I. When the holder has cried down, or when the referee has cried down, or when the umpire has called foul.

II. When a goal has been obtained.

III. When it has gone into touch, or touch-in-goal, except for punt-out.

IV. When a touch-down or safety has been made.

V. When a fair catch has been heeled. No play can be made while the ball is dead, except to put in play by rule.

RULE 12. The grounds must be 330 feet in length and 160 feet in width, with a goal place in the middle of each goal line, composed of two upright posts, exceeding 20 feet in height, and placed 18 feet 6 inches apart, with cross-bar 10 feet from the ground.

RULE 13. The game shall be played by teams of eleven men each, and in case of a disqualified or injured player a substitute shall take his place. Nor shall the disqualified or injured player return to further participation in the game.

Amendment adopted at a special meeting of the Intercollegiate Association, 1893: " No member of a graduate department, nor a special student shall be allowed to play, nor any undergraduate who has registered or attended lectures or recitations at any other university or college nor by any undergraduate who is not pursuing a course requiring for a degree an attendance of at least three years."

RULE 14. There shall be an umpire and a referee. No man shall act as an umpire who is an alumnus of either of the competing colleges. The umpires shall be nominated and elected by the Advisory Committee. The referee shall be chosen by the two captains of the opposing teams in each game, except in case of disagreement, when the choice shall be referred to the Advisory Committee, whose decision shall be final. All the referees and umpires shall be permanently elected and assigned on or before the third Saturday in October in each year.

RULE 15. (*a*) The umpire is the judge for the players, and his decision is final regarding fouls and unfair tactics.

(*b*) The referee is judge for the ball, and his decision is final in all points not covered by the umpire.

(*c*) Both umpire and referee shall use whistles to indicate cessation of play on fouls and downs. The referee shall use a stop-watch in timing the game.

(*d*) The umpire shall permit no coaching, either by substitutes, coaches, or any one inside the ropes. If such coaching occur he shall warn the offender, and upon the second offense must have him sent behind the ropes for the remainder of the game.

RULE 16. (*a*) The time of a game is an hour and a half, each side playing forty-five minutes from each goal. There shall be ten minutes' intermission between the two halves. The game shall be decided by the score of even halves. Either side refusing to play after ordered to by the referee, shall forfeit the game. This shall also apply to refusing to commence the game when ordered to by the referee. The referee shall notify the captains of the time remaining, not more than ten, nor less than five, minutes from the end of each half.

(*b*) Time shall not be called for the end of a three-quarter until the ball is dead; and in the case of a try-at-goal from a touch-down the try shall be allowed. Time shall be taken out while the ball is being brought out, either for a try, kick-out, or kick-off.

RULE 17. No one wearing projecting nails or iron plates on his shoes, or any metal substance upon his person, shall be allowed to play in a match. No sticky or greasy substance shall be used on the person of players.

RULE 18. The ball goes into touch when it crosses the side line, or when the holder puts part of either foot across or on that line. The touch line is in touch, and the goal line in goal.

RULE 19. The captains shall toss up before the commencement of the match, and the winner of the toss shall have his choice of goal or of kick-off. The same side shall not kick off in two successive halves.

RULE 20. The ball shall be kicked off at the beginning of each half; and whenever a goal has been obtained, the side which has lost it shall kick off. (See Rules 32 and 34.)

RULE 21. A player who has made and claimed a fair catch shall take a drop-kick, or a punt, or place the ball for a place-kick. The opponents may come up to the catcher's mark, and the ball must be kicked from some spot behind that mark on a parallel to touch line.

RULE 22. The side which has a free-kick must be behind the ball when it is kicked. At kick-off the opposite side must stand at least ten yards in front of the ball until it is kicked.

RULE 23. Charging is lawful for opponents if a punter advances beyond his line, or in case of a place-kick, immediately the ball is put in play by touching the ground. In case of a punt-out, not till ball is kicked.

RULE 24. (*a*) A player is put off side, if, during a scrimmage he gets in front of the ball, or if the ball has been last touched by his own side behind him. It is impossible for a player to be off side in his own goal. No player when off side shall touch the ball, or interrupt, or obstruct opponent with his hands or arms until again on side.

(*b*) A player being off side is put on side when the ball has touched an opponent, or when one of his own side has run in front of him, either with the ball, or having touched it when behind him.

(*c*) If a player when off side touches the ball inside the opponents' five-yard line, the ball shall go as a touch-back to the opponents.

RULE 25. No player shall lay his hands upon, or interfere by use of hands or arms, with an opponent, unless he has the ball. The side which has the ball can only interfere with the body. The side which has not the ball can use the hands and arms, as heretofore.

RULE 26. (*a*) A foul shall be granted for intentional delay of game, off side play, or holding an opponent, unless he has the ball. No delay arising from any cause whatsoever shall continue more than five minutes.

(*b*) The penalty for fouls and violation of rules, except otherwise provided, shall be a down for the other side; or, if the side making the foul has not the ball, five yards to the opponents.

Rule 27. (*a*) A player shall be disqualified for unnecessary roughness, hacking or striking with closed fist.

(*b*). For the offenses of throttling, tripping up or intentional tackling below the knees, the opponents shall receive twenty-five yards, or a free-kick, at their option. In case, however, the twenty-five yards would carry the ball across the goal line they can have half the distance from the spot of the offense to the goal line, and shall not be allowed a free-kick.

Rule 28. A player may throw or pass the ball in any direction except towards opponents' goal. If the ball be batted in any direction or thrown forward it shall go down on the spot to opponents.

Rule 29. If a player when off side interferes with an opponent trying for a fair catch, by touching him or the ball, or waving his hat or hands, the opponent may have a free-kick, or down, where the interference occurred.

Rule 30. (*a*) If a player having the ball be tackled and the ball fairly held, the man so tackling shall cry "held," the one so tackled must cry "down," and some player of his side put it down for a scrimmage. The snapper back and the man opposite him cannot pick out the ball with the hand until it touch a third man; nor can the opponents interfere with the snapper-back by touching the ball until it is actually put in play. Infringement of this nature shall give the side having the ball five yards at every such offense. The snapper-back is entitled to full and undisturbed possession of the ball. If the snapper-back be off side in the act of snapping back, the ball must be snapped again; and if this occurs three times on same down, the ball goes to opponents. The man who first receives the ball, when snapped back from

18

a down, or thrown back from a fair, shall not carry the ball forward under any circumstances whatever. If, in three consecutive fairs and downs, unless the ball cross the goal line, a team shall not have advanced the ball five or taken it back twenty yards, it shall go to the opponents on spot of fourth. "Consecutive" means without leaving the hands of the side holding it, and by a kick giving opponents fair and equal chance of gaining possession of it. When the referee, or umpire, has given a side five yards, the following down shall be counted the first down.

(*b*) The man who puts the ball in play in a scrimmage cannot pick it up until it has touched some third man. "Third man" means any other player than the one putting the ball in play and the man opposite him.

RULE 31. If the ball goes into touch, whether it bounds back or not, a player on the side which touches it down must bring it to the spot where the line was crossed, and there either

I. Bound the ball in the field of play or touch it in with both hands at right angles to the touch line, and then run with it, kick it, or throw it back; or

II. Throw it out at right angles to the touch line; or

III. Walk out with it at right angles to touch line any distance not less than five nor more than fifteen yards, and there put it down, first declaring how far he intends walking. The man who puts the ball in must face field or opponents' goal, and he alone can have his foot outside touch line. Any one except him who puts his hands or feet beween the ball and his opponents' goal is off side. If it be not thrown out at right angles either side may claim it thrown over again, and if it fail to be put in play fairly in three trials it shall go to the opponents.

RULE 32. A side which has made a touchdown in their opponents' goal *must* try at goal, either by a place-kick or a punt-out. If the goal be missed the ball shall go as a kick-off at the center of the field to the defenders of the goal.

RULE 33. (*a*) If the try be by a place-kick, a player of the side which has touched the ball down shall bring it up to the goal line, and, making a mark opposite the spot where it was touched down, bring it out at right angles to the goal line such distance as he thinks proper, and there place it for another of his side to kick. The opponents must remain behind their goal line until the ball has been placed on the ground.

(*b*) The placer in a try-at-goal may be off side or in touch without vitiating the kick.

RULE 34. If the try be by a punt-out the punter shall bring the ball up to the goal line, and, making a mark opposite the spot where it was touched down, punt out from any spot behind line of goal and not nearer the goal post than such mark, to another of his side, all of whom must stand outside of goal line not less than fifteen feet. If the touchdown was made in touch-in-goal the punt-out shall be made from the intersection of the goal and touch lines. The opponents may line up anywhere on the goal line except space of five feet on each side of punter's mark, but cannot interfere with punter, nor can he touch the ball after kicking it until it touch some other player. If a fair catch be made from a punt-out the mark shall serve to determine positions as the mark of any fair catch. If a fair catch be not made on the first attempt the ball shall be punted over again, and if a fair catch be not made on the second attempt the ball shall go as a kick-off at the center of the field to the defenders of the goal.

RULE 35. A side which has made a touch back or a safety must kick out, except as otherwise provided (see rule 32), from not more than twenty-five yards outside the kicker's goal. If the ball go into touch before striking a player it must be kicked out again, and if this occurs three times in succession it shall be given to opponents as in touch on twenty-five-yard line on side where it went out. At kick-out opponents must be on twenty-five-yard line or nearer their own goal.

RULE 36. The following shall be the value of each point in the scoring:

Goal obtained by touchdown, - - -	6
Goal from field kick, - - - - -	5
Touchdown failing goal, - - - -	4
Safety by opponents. - - - - -	2

www.ingramcontent.com/pod-product-compliance
Lightning Source LLC
Chambersburg PA
CBHW021515210326
41599CB00012B/1263